Paulina Pianko-Oprych

Cell, Stack and System Modelling

Paulina Pianko-Oprych

Cell, Stack and System Modelling

Solid Oxide Fuel Cell

LAP LAMBERT Academic Publishing

Impressum / Imprint

Bibliografische Information der Deutschen Nationalbibliothek: Die Deutsche Nationalbibliothek verzeichnet diese Publikation in der Deutschen Nationalbibliografie; detaillierte bibliografische Daten sind im Internet über http://dnb.d-nb.de abrufbar.

Alle in diesem Buch genannten Marken und Produktnamen unterliegen warenzeichen-, marken- oder patentrechtlichem Schutz bzw. sind Warenzeichen oder eingetragene Warenzeichen der jeweiligen Inhaber. Die Wiedergabe von Marken, Produktnamen, Gebrauchsnamen, Handelsnamen, Warenbezeichnungen u.s.w. in diesem Werk berechtigt auch ohne besondere Kennzeichnung nicht zu der Annahme, dass solche Namen im Sinne der Warenzeichen- und Markenschutzgesetzgebung als frei zu betrachten wären und daher von jedermann benutzt werden dürften.

Bibliographic information published by the Deutsche Nationalbibliothek: The Deutsche Nationalbibliothek lists this publication in the Deutsche Nationalbibliografie; detailed bibliographic data are available in the Internet at http://dnb.d-nb.de.

Any brand names and product names mentioned in this book are subject to trademark, brand or patent protection and are trademarks or registered trademarks of their respective holders. The use of brand names, product names, common names, trade names, product descriptions etc. even without a particular marking in this works is in no way to be construed to mean that such names may be regarded as unrestricted in respect of trademark and brand protection legislation and could thus be used by anyone.

Coverbild / Cover image: www.ingimage.com

Verlag / Publisher:
LAP LAMBERT Academic Publishing
ist ein Imprint der / is a trademark of
OmniScriptum GmbH & Co. KG
Heinrich-Böcking-Str. 6-8, 66121 Saarbrücken, Deutschland / Germany
Email: info@lap-publishing.com

Herstellung: siehe letzte Seite /
Printed at: see last page
ISBN: 978-3-659-62295-3

Table of contents

Nomenclature

c	constant volume specific heat, J/kg
c_0	constant volume specific heat, J/kg
d	particle diameter, m
d_p	pore diameter, m
$D_{m,j}$	diffusion coefficient of species m, m^2/s
D_{mn}	binary diffusion coefficient of species m in species n, m^2/s
D_{mK}	Knudsen diffusion coefficient of species m, m^2/s
e	specific internal energy, J/kgK
E	electromotive force, V
E_{Act}	activation energy of electrochemical reaction, J/kmol
$E_{act,a}$	anodic activation energy, J/kmol
$E_{act,c}$	cathodic activation energy, J/kmol
E_{aux}	auxiliary power consumption, W
E_{Nernst}	Nernst potential of cell, V
E_0	standard cell potential under standard conditions, V
F	Faraday constant, C/mol
$F_{h,j}$	thermal flux, J/m^2s
$F_{m,j}$	diffusive mass flux of species m, $kg/m^2{\cdot}s$
$h_{t,m}$	specific enthalpy of species m, J/kg
i	current density, A/m^2
i_{elec}	electron transfer in cell electrolyte, $C/m^3{\cdot}s$
i_{ion}	ion transfer in cell electrolyte, $C/m^3{\cdot}s$
I	transfer current, $C/m^3{\cdot}s$
I_{cell}	cell current, A
I_e	mean current density in electrolyte, $C/m^2{\cdot}s$
I_0	exchange current density, $C/m^2{\cdot}s$
K	thermal conductivity of gas mixture or solid materials, W/m·K
K_i	permeability, kg/m^3s
l_e	electrolyte thickness, m
m	number of species, -
$m_{H_2,cons}$	mole flow rate of hydrogen consumed, kmol/s
$m_{H_2,in}$	mole flow rate of hydrogen fuel input to SOFC, kmol/s

M_m	molecular weight of species m, kg/kmol
M_n	molecular weight of species n, kg/kmol
n	number of electrons, -
n_{fuel}	molar fuel flow, kmol/s
N_A	Avogadro number, 1/kmol
N_{fuel}	mass flow of fuel, kmol/s
p	pressure, Pa
p_{H_2a}	partial pressure of H_2 at the electrode-electrolyte interface, Pa
$p_{H_2O,a}$	partial pressure of H_2O at the electrode-electrolyte interface, Pa
p_{O_2a}	partial pressure of O_2 at the electrode-electrolyte interface, Pa
p_i^*	partial pressure of the i-th reactant/product at the reaction sites, Pa
p_i^0	partial pressure of the i-th reactant/product, Pa
$P_{out,gross}$	gross electrical power output of SOFC, W
$Q_{heat-recovery}$	cogeneration heat rate, W
q_{rad}	radiative heat flux, $J/m^2 \cdot s$
R	gas constant, $J/kmol \cdot K$
R_e	cell reaction rate, $mole/m^3 \cdot s$
s_c	chemical energy source due to endo- or exothermic chemical processes, $J/m^3 \cdot s$
s_e	heat source due to electrochemical processes, $J/m^3 \cdot s$
s_h	energy source due to ohmic resistance and radiation, $J/m^3 \cdot s$
s_i	i-th momentum source component, $kg/m^2 \cdot s^2$
s_m	mass production/consumption rate, $kg/m^3 \cdot s$
s_Σ	mass source, $kg/m^3 \cdot s$
T	temperature, K or °C
T_{surf}	MEA surface temperature, K or °C
T_∞	furnace temperature, K or °C
u_i	velocity component, m/s
U_f	fuel utilisation factor, %
U_i	superficial velocity, m/s
V_{cell}	cell voltage, V
$\dot{V}_i$	volumetric flow rate of the i-component, m^3/s

W_{el}	electric work, J/kg
x_i	i-th direction, -
x_{O_2}	volume fraction of oxygen, m^3/m^3
X_{CH_4}	methane contribution to hydrogen formation, -
X_{CO}	carbon monoxide contribution to hydrogen formation, -
Y_m	mass fraction of species m, kg/kg

Greek letters

α_i	i-th coefficient, -
β	symmetry coefficient, -
β_i	i-th coefficient, -
Γ_{H_2}	fuel equivalent hydrogen content, -
γ_a	anodic pre-exponential coefficient, -
γ_c	cathodic pre-exponential coefficient, -
ε	porosity, -
η_{act}	activation polarisations, V
η_{conc}	concentration polarisation, V
$\eta_{DC/AC}$	efficiency of DC/AC conversion, -
η_{el}	electrical efficiency, %
η_{Ohm}	ohmic polarisation, V
η_{stack}	stack efficiency, %
η_{system}	system efficiency, %
$\eta_{thermal}$	thermal efficiency, %
λ	air to fuel ratio, -
ξ	MEA surface emissivity, -
μ	molecular dynamic viscosity of the gas mixture, kg/m·s
μ	chemical potential, V
μ^0	standard chemical potential, V
ρ	gas mixture density, kg/m^3
σ	electronic/ionic conductivity, $1/\Omega \cdot m$

σ_e	electrolyte anionic conductivity, $1/\Omega \cdot m$
σ_{elec}	electronic conductivity in anode, $1/\Omega \cdot m$
σ_{ion}	ion conductivity in electrolyte material, $1/\Omega \cdot m$
σ_{mn}	collision diameter Lennard-Jones potential model, m
σ_{SB}	Stefan-Boltzmann constant, $W/m^2 \cdot K^4$
Ω_{mn}	collision dimensionless Lennard-Jones integral, -
δ	thickness, m
δ_{ij}	Kronecker delta, -
τ	tortuosity, -
τ_{ij}	stress tensor, Pa
ϕ	potential, V
ϕ_{elec}	electron potential, V
ϕ_{ion}	ion potential, V
$\Phi_{e,int-a}$	electric potential of the outermost layer of anode interconnects, V
$\Phi_{e,int-c}$	electric potential of the outermost layer of cathode interconnects, V
$\overline{\Delta h_f}$	enthalpy of formation, J/kmol
ΔH_e	enthalpy change, J/kmol
$\overline{\Delta g_f}$	Gibbs free energy, J/kmol
ΔG^0	standard Gibbs free energy change of cell reaction, J/kmol

Subscripts

a	anode
act	activation polarisation
c	cathode
cons	consumed
e	electrode (anode, cathode)
elec	electrolyte, electrons
f	fuel
ion	ions
i, j, k	Cartesian directions
m, n	species index

Acronyms

ADL   Anode Diffusion Layer,

APUs   Auxiliary Power Units,

ATR   Autothermal Reforming,

CFD   Computational Fluid Dynamics,

CHP   Combined Heat and Power system,

CPO_x Catalytic Partial Oxidation,

FC   Fuel Cell,

FEM   Finite Element Method,

HHV   fuel High Heating Value, J/kmol

LHV   fuel Lower Heating Value, J/kmol

LSM   strontium-doped lanthanum manganite,

MEA   Membrane-Electrode Assembly

O/C   oxygen to carbon ratio,

OCV   Open Circuit Voltage,

S/C   steam to carbon volume ratio,

SOFC   Solid Oxide Fuel Cell,

SR   Steam Reforming,

YSZ   Yttria Stabilized Zirconia.

1. Introduction

Simulations of systems based on SOFC for generating electrical energy are widely described in [1-3]. In the first part of Introduction basic definitions are presented, then multiscale modelling approach is discussed.

1.1. Basic definitions

Solid Oxide Fuel Cell (SOFC) technology is receiving increasing scientific and industrial interest for a number of stationary and portable applications, such as Combined Heat and Power (CHP) systems [4] and Auxiliary Power Units (APUs) [5]. Certainly, high electrical efficiency and long-term durability of a SOFC are the key requirements for the introduction of the fuel cell technology into the market. However, there are several different parameter sets that exert a significant impact on electrical efficiency and power of SOFC systems. Air utilisation factor, air to fuel ratio or fuel utilisation factor are only a few of the many other factors that should be considered.

The cell electrical efficiency, η_{el}, can be defined as a function of the cell voltage, V_c, fuel utilisation factor, U_f, and inlet composition, Γ_{H_2}, according to equation (1.1) proposed by [6]:

$$\eta_{el} = \frac{W_{el}}{LHV} = \frac{n \cdot F \cdot E}{LHV} = \frac{n \cdot F \cdot V_c + U_f + \Gamma_{H_2}}{LHV} \qquad (1.1)$$

where: W_{el} is the electrical energy produced per mole of fuel, LHV is known as the Lower Heating Value and is equal to the value of the enthalpy of formation, $\overline{\Delta h_f}$, that would be produced by burning fuel and formation of water vapours. Its value is negative when energy is released (LHV = - $\overline{\Delta h_f}$) [7].

The cell electrical efficiency can be also calculated according to equation (1.2) proposed by [8]:

$$\eta_{el} = \frac{Power}{n_{fuel} \cdot LHV} = \frac{I_{cell} \cdot V_{cell}}{n_{fuel} \cdot LHV} \qquad (1.2)$$

where the molar fuel flow, n_{fuel}, is defined as follows:

$$n_{fuel} = \frac{n_{H_2 equivalent}}{n_{H_2 in} + n_{CO,in} + 4 \cdot n_{CH_4,in} + 7 \cdot n_{C_2H_6,in} + 10 \cdot n_{C_3H_8,in} + 13 \cdot n_{C_4H_{10},in}} \qquad (1.3)$$

where $n_{H_2 equivalent}$ denotes the equivalent hydrogen flow rate:

$$n_{H_2 equivalent} = \frac{n_{H_2 cons}}{U_f} = \frac{I}{2F \cdot U_f} \qquad (1.4)$$

The fuel utilisation factor, U_f, is defined as the ratio of the molar flow rate of hydrogen consumed, $m_{H_2,cons}$, to the molar flow rate of hydrogen fuel input to a SOFC, $m_{H_2,in}$, (equation (1.5)):

$$U_f = \frac{m_{H_2,cons}}{m_{H_2,in}} = \frac{m_{H_2,in} - m_{H_2,out}}{m_{H_2,in}} \qquad (1.5)$$

The fuel equivalent hydrogen content, Γ_{H_2}, is expressed for a fuel mixture of H_2-CH_4-CO-H_2O by equation (1.6):

$$\Gamma_{H_2} = X_{H_2} + X_{CH_4} + X_{CO} \qquad (1.6)$$

where: X_{CH_4} and X_{CO} are methane and carbon monoxide contributions to hydrogen formation, accounting for a complete hydrocarbon reforming and for the additional contribution of CO shift reaction.

The maximum electrical energy available is equal to the ratio of the change in Gibbs free energy, $\overline{\Delta g_f}$, to the enthalpy of formation, $\overline{\Delta h_f}$, and can be defined by equation (1.7):

$$\eta_{max} = \eta_{therm} = \frac{\overline{\Delta g_f}}{\Delta h_f} \cdot 100\% \tag{1.7}$$

Thus, the overall electrical efficiency, $\eta_{electrical}$, can be generally described as:

$$\eta_{electrical} = \frac{electrical\ work\ output}{chemical\ energy\ input} = \eta_{therm} \cdot \eta_{el} \cdot U_f \cdot \eta_{reformer} \cdot \eta_{systemcomponents} \tag{1.8}$$

Similarly, the air utilisation factor, U_a, is proportional to the oxygen quantity extracted from the air flow to oxidise fuel and can be defined as follows (equation (1.9)):

$$U_a = \frac{m_{O_2,cons}}{m_{O_2,in}} = \frac{m_{O_2,in} - m_{O_2,out}}{m_{O_2,in}} \tag{1.9}$$

The air to fuel ratio, R_f, is equal to the number of moles of air required to oxidise 1 mole of the given fuel and air utilisation factors and can be calculated from equation (1.10) [6]:

$$R_f = \frac{m_{air,inlet}}{m_{fuel,requested}} \tag{1.10}$$

To achieve high efficiency of a SOFC, high fuel utilisation is required, while fuel utilisation results from high concentration gradients at the anode, where the fuels like H_2, CO, CH_4 are successively diluted by reaction products, such as H_2O, CO_2. Gas concentrations at the cell and stack level continuously vary along the flow path leading to temperature and local power density changes. These variations affect both efficiency through a large concentration polarisation and long-term stability through thermo-mechanical stress and local operating conditions, which may lead to SOFC degradation [9]. The operation procedure for heat up and shut down has to be considered for proper stack temperature control. Furthermore, soot formation has to be analysed to provide long term durability of a system [4]. Also the integration of heat exchangers, fuel processor and an off-gas burner is a necessity at the system level for end users. The system performance can be evaluated based on the net electrical efficiency, η_{net}, and the overall system efficiency, η_{system}, as shown in [10]:

$$\eta_{net} = \frac{P_{out,gross} \cdot \eta_{DC/AC} - E_{aux}}{N_{fuel} \cdot LHV} \qquad (1.11)$$

$$\eta_{system} = \frac{(P_{out,gross} \cdot \eta_{DC/AC} - E_{aux}) + Q_{heat-recovery}}{N_{fuel} \cdot LHV} \qquad (1.12)$$

where: $P_{out,gross}$ is the gross electrical power output of a SOFC, $\eta_{DC/AC}$, is the efficiency of DC/AC conversion, E_{aux}, is the auxiliary power consumption, $Q_{heat-recovery}$ is the cogeneration heat rate, N_{fuel} is the mass flow of fuel.

To give a more standardised means of comparing efficiencies, the following definitions of efficiency are often used [11]:

$$\eta_{stack} = \frac{DC \ electricity \ output}{HHV \ of \ hydrogen \ energy \ input} \qquad (1.13)$$

$$\eta_{system} = \frac{AC \ electricity \ output - electricity \ consumption}{HHV \ of \ natural \ gas \ input} \qquad (1.14)$$

$$\eta_{thermal} = \eta_{system} + \frac{Useful \ heat \ output \ recovered \ in \ coolant \ water}{HHV \ of \ natural \ gas \ input \ to \ reformer \ aux.burners} \qquad (1.15)$$

An extensive comparison of the efficiency of SOFC systems tested from 2000 to 2008 was reported by [11].

1.2. Multi-scale modelling

In general, understanding the SOFC system performance at different length scales: system, component and fluid transport at the micro- and macrostructure of electrodes up to cell scale is one of the main issues of further SOFC system development and commercialisation. Development of a SOFC system requires proper modelling approaches and the use of numerical process simulators, which will provide clear insight into various aspects of the system operation. One way to couple mass and heat transport phenomena with electrochemical processes at the micro-scale with velocity and temperature distributions in the air and fuel channels at the macro-scale while including aspects of system components integration is to use multi-scale modelling approach in fuel cell research.

Multi-scale modelling applied generally to chemical engineering and more specifically to SOFC fuel cells was specifically presented by [12] and [13], respectively. Multi-scale approach covers three main types of modelling methods: (i) computational chemistry, (ii) CFD methods and (iii) process simulator tools. Relation between length and time scales with modelling methods is presented in Figure 1.1.

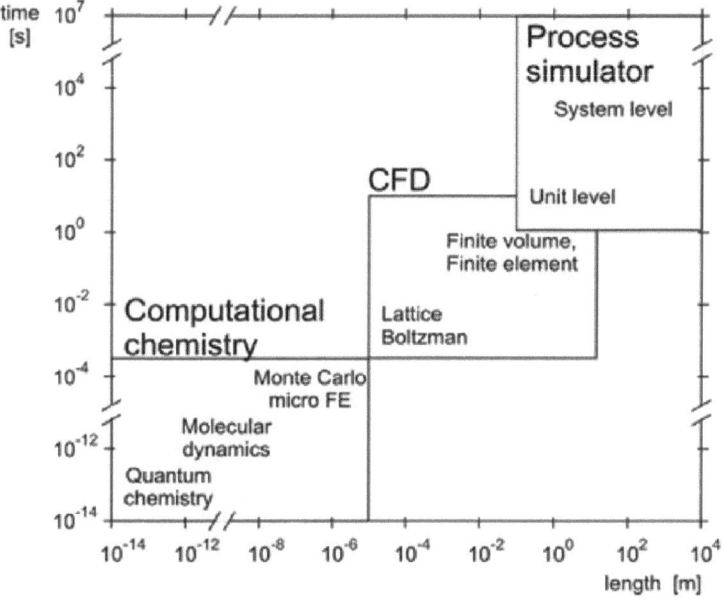

Figure 1.1. Computational tools in multi-scale modelling according to [12]

Andersson et al. [14] was one of the first to present a division of the modelling methods into the SOFC multi-scale modelling using different length and time scales as a

11

criterion and highlighting the following groups of methods: micro-, meso- and macro-scales. The micro-scale model corresponds to the atom or molecular level as thermo- or fluid dynamics and detailed chemical reactions are considered. Density Functional Theory (DT), Quantum Chemistry (QC), Lattice-Boltzmann method (LBM), Molecular Dynamics (MD) and Mechanistic Models (MM) are included by [14] in the first group, while, the Monte Carlo (MC), Brownian Dynamics (BD) and Dissipative Particle Dynamics (DPD) methods correspond to the meso-scale. This group of methods based on models, which correspond to a scale larger than a particle, but smaller than the facility or global flow field. Finally, the last group of methods related to the macro-scale includes: Finite Element Method (FEM), Finite Volume Method (FVM), Finite Difference Method (FDM) and Spectral Methods (SM). These methods describe the global flow field [14].

Grew and Chiu [15] proposed allocation of computational methods not only with regard to time and length scales, but including functional and degradation processes in the heterogeneous SOFC electrode structures as can be seen in Figure 1.2.

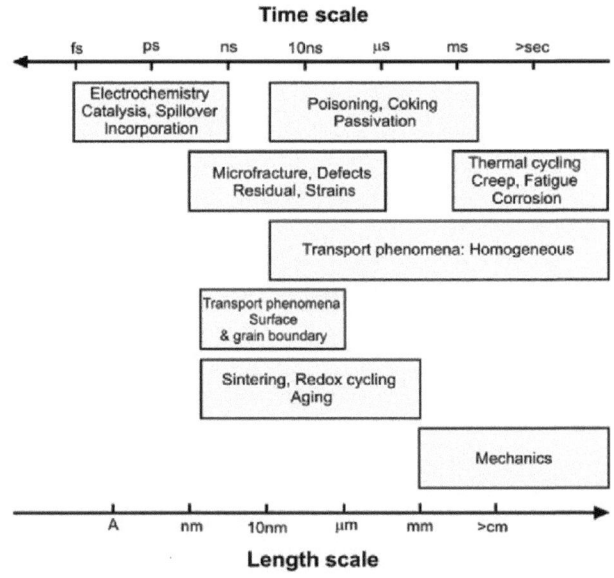

Figure 1.2. Different process in SOFC and corresponding length and time scales [15]

The role of computation in supporting SOFC system development was clearly recognised by [16], who proposed a simulation tool to predict the performance and the main electro-thermal-fluid dynamic characteristics of an integrated biomass fuelled gasifier SOFC system. The gasifier was modelled by 0D model capable to provide output starting from any biomass composition and water content, while the SOFC behaviour was predicted by a 3D CFD – SOFC module, which included gas evolution into gas channels and porous electrodes. The internal reforming was modelled within Cordiner's et al. [16] approach by the introduction of the homogeneous phase methane reforming and Water-Gas-Shift (WGS) reactions. The system performance was analysed using a specifically developed numerical model, which enables implementation of both gasifier and SOFC modules. The numerical model delivered a quantitative data about current distribution on the electrolyte and about the main species moving in the gas channels and electrodes at the good system performance of 45.8% efficiency [16].

Bessler [13] built on previous efforts and developed a comprehensive approach of mathematical modelling of SOFC schematically presented in Table 1.1, that includes: (i) diffusion through the surface, (ii) laminar flow through porous media at the electrode level, (iii) transport of oxidant and fuel at the fuel cell level, (iv) electrical circuits of the cell and thermal balancing, (v) integration of the entire system with balance-of-plant components.

At the surface level Bessler [13] estimated the elementary kinetic and thermodynamic properties using the CANTERA software, while at the electrode and fuel cell levels he used the in-house simulation software DENIS (Detailed Electrochemistry and Numerical Impedance Simulation). Both these numerical codes were coupled to the system simulation software CFD - ANSYS at the fuel cell level and Matlab-SIMULINK at the system level. Depending on type of modelling and desired complexity or level of details, sufficient data have to be fed into the model. In particular, at the surface level, when kinetic models were used, thermodynamic parameters such as molar enthalpies and entropies of all intermediates as well as kinetic coefficients of reactions and surface or bulk diffusion coefficients were necessary to be delivered to the code. Each model was able to predict behaviour of a particular system level and to provide information on the optimal operating parameters. Thereby, crucial data for the system improvements were delivered [13].

Table 1.1. Multi-scale modelling approach of chemical and transport processes in a SOFC [13]

Level	Length & time scales	Geometry	Chemistry	Transport	Modelling approach
System	10^0 m 10^4 s			Mass, energy fluxes	Process simulators: Aspen Plus™, HYSYS, Matlab-SIMULINK
Stack	10^{-1} m 10^2 s			Laminar flow Heat conduction radiation	CFD codes: ANSYS-Fluent, COMSOL
Fuel Cell	10^{-2} m 10^0 s		Gas-phase chemistry	Laminar flow Heat convection	Heat, mass and charge transport with DENIS
Electrode	10^{-4} m 10^{-2} s		Thermal and redox cycling	Porous multiphase mass and charge transport	Continuum transport and percolation theory with DENIS
Surface	10^{-8} m 10^{-6} s		Surface chemistry	Surface diffusion	Mean field elementary kinetics with CANTERA

2. Cell level modeling

A number of modelling studies with various degrees of complexity have been carried out to determine the factors that influence electrode properties and to predict electrode performance [17 - 19]. Significant effort has been made to develop a deeper understanding of the mechanisms of processes taking place in the Solid Oxide Fuel Cells [20 - 21].

2.1. Micro-modelling level

The electrochemical performance models for SOFC electrodes are defined with simple empirical lumped-parameter kinetic models such as the Butler-Volmer equation to highly detailed micro-kinetic models that are based on finite-element analysis and consider microstructure, material properties, elementary kinetics and transport phenomena [20]. Recent advances in 3D imaging using Scanning Electron Microscopy (SEM) images [19], FIB-SEM [22] or X-ray microtomography [23 - 25] make it possible to represent the microstructure of an electrode at the submicron scale. Such images are used to create a model geometry that reproduces a sample of the microstructure. Kreller et al. [20] distinguished three general classes of microstructural representations that are commonly employed in micro-kinetic modelling of porous SOFC electrodes as is shown in Figure 2.1. In the first class (macro-homogeneous approach) all characteristics of a microstructure are represented in terms of volume-average macro-homogeneous parameters such as porosity or transport tortuosity [26]. Such models can predict well when the active region of an electrode is large in comparison to individual microstructural features, such as particle dimensions [27 - 28].

In the second class, microstructure is represented by pseudo-particles, which have a size and arrangement chosen to match a sub-set of average microstructural properties such as porosity and surface area. A pseudo-particle model is able to capture the effect of 3D gradients near the three-phase boundary, when the size of the utilisation region is competitive with the microstructure [20]. However, this approach does not take into account many other details of the microstructure, such as particle size dispersion, interfacial contact angle or interfacial contact area.

15

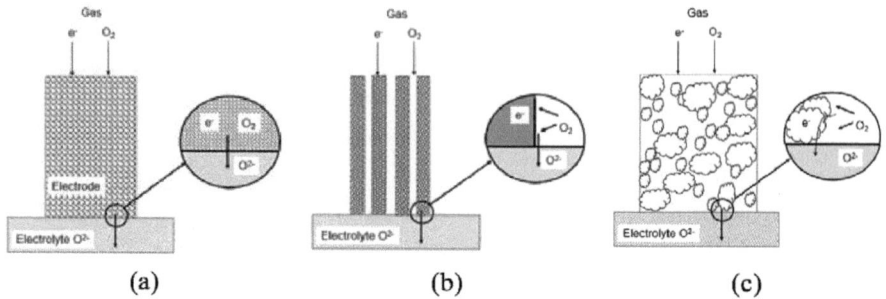

Figure 2.1. Various approximations of a microstructure used in micro-kinetic modelling of porous SOFC electrodes: a) macro-homogeneous approach, b) pseudo particles, c) 3D geometry based on images of microstructures [20]

In the third approach, an experimentally characterised sample of a 3D microstructure based on FIB-SEM reconstruction is used as a model for the microstructure as a whole. These models accurately represent all the important aspects of a microstructure for its kinetics, but simulation times are very long [20].

Kreller et al. [20] tried to answer the question of which of the three approaches is the best one. They simulated linear impedance and non-linear impedance of single-phase mixed-conducting cathodes comprising a variety of model geometries. The team also examined hybrid models performing detailed simulations near the three-phase boundary and the boundary was incorporated into the effective media approach farther away from the electrode-electrolyte interface. Their predictions based on cylindrical pseudo-particles and analytical solutions were nearly identical to full simulation data suggesting that a relatively crude microstructural model represented by monodispersed cylinders is capable to predict most features of linear impedance of a real microstructure. This also means that electrode performance can be accurately predicted using purely volume-averaged properties based on the macro-homogeneous pseudo-particle models. However, when Kreller et al. [20] extended simulations to the second and third harmonic it turned out that the non-linear impedance spectra of various microstructures were very sensitive to the details of the microstructure near the interface.

The incorporation of elementary heterogeneous chemical kinetics in the form of a multistep reaction mechanism was emphasised by [29]. The model included the following five elementary reactions in the Ni-YSZ three phase region for the electrochemical oxidation of hydrogen:

1. Adsorption/desorption on the Ni surface (eq. (2.1)):

$$H_{2(g)} + 2(Ni) \rightarrow 2H(Ni) \tag{2.1}$$

2. Two charge-transfer reactions at the TPB region (eq. (2.2)-(2.3)):

$$H(Ni) + O^{2-}(YSZ) \rightarrow (Ni) + OH^-(YSZ) + e^-(Ni) \tag{2.2}$$

$$H(Ni) + OH^-(YSZ) \rightarrow (Ni) + H_2O(YSZ) + e^-(Ni) \tag{2.3}$$

3. Adsorption/desorption on the YSZ surface (eq. (2.4)):

$$H_2O(YSZ) \rightarrow H_2O_{(g)} + (YSZ) \tag{2.4}$$

4. Transfer of oxygen ions between the surface and the bulk YSZ (eq. (2.5)):

$$O_O^X(YSZ) + (YSZ) \rightarrow O^{2-}(YSZ) + V_O^{''}(YSZ) \tag{2.5}$$

where H(Ni) denoted adsorbed atomic hydrogen on the Ni anode surface, (Ni) was the empty surface site, e^-(Ni) was an electron within the Ni anode. For the YSZ electrolyte, $O_O^X(YSZ)$ was a lattice oxygen and $V_O^{''}(YSZ)$ was an oxygen vacancy. In addition, three species: $O^{2-}(YSZ)$, $OH^-(YSZ)$, $H_2O(YSZ)$ and empty sites (YSZ) were proposed at the YSZ surface [29].

The overall oxygen reduction at the electrode-electrolyte interface was defined in two steps:

1. Adsorption/desorption (eq. (2.6)):

17

$$O_{2(g)} + 2(c) \rightarrow 2O_{ad}(c) \qquad\qquad (2.6)$$

2. Charge transfer and incorporation at the TPB (eq. (2.7)):

$$O_{ad}(c) + V_O^{''}(el) + 2e^-(c) \rightarrow O_O^x(el) + (c) \qquad\qquad (2.7)$$

where: $O_{ad}(c)$ was adsorbed atomic oxygen at the cathode surface, (c) was the unoccupied cathode surface site, $V_O^{''}(el)$ and $O_O^x(el)$ denote respectively oxygen vacancies and lattice oxygen ions in the bulk of electrolyte, $e^-(c)$ represents an electron within the cathode. This approach assumed a weak coupling between thermal heterogeneous chemistry within the bulk of the porous anode and charge-transfer chemistry in the relatively thin three-phase region. The charge-transfer chemistry proceeded on the basis of hydrogen concentration at the interface between the anode and the dense electrolyte. The proposed approach by [29] neglected effects such as any charge-transfer inhibition associated with other adsorbed species competing with the adsorbed H(Ni). Nevertheless, this approach represented a significant advance in the fuel cell modelling.

In the subject literature, research results were published on developing fully coupled elementary, thermal and electrochemical reaction mechanisms [30]. Although Janardhanan and Deutschmann [30] employed a similar approach to that of Zhu's et al. [29], they did not make isothermal assumptions for a planar anode-supported cell fuelled by humidified methane. Both steam and dry reforming reactions were considered. Drops in temperature along the fuel cell length in the membrane-electrode assembly (MEA) were found near the inlets for the co-flow configuration due to the endothermic reforming reactions. Janardhanan et al. [31] developed a mathematical model based on porosity, particle diameter and volume fraction of the ionic and electronic phase. They concluded that the maximum volume specific TPB length was achieved at 50% porosity.

Vogler et al. [32] considered the electrochemical hydrogen oxidation reaction at Ni/YSZ anodes based on equations (2.1) - (2.5), the YSZ surface reaction between water and oxygen ions, the Ni surface reaction for water and hydroxyl ions as well as five additional charge transfer equations. Their model was based on elementary physical and chemical processes without the assumption of a specific rate-determining step. The hydrogen spillover mechanism described by equation (2.3) was found to provide the best agreement with their experimental data.

Heterogeneous electrode properties at the micro-scale level were also examined by [33], who predicted overpotential losses by capturing the coupled mass transfer and electrochemical reactions. Shi and Xue [34] introduced a concept including polarisation effects, molecular and Knudsen diffusion, gas transport, ionic and electronic conduction, surface diffusion of intermediate species and electrochemical reactions. Their results indicated that cell performance was strongly dependent on porous microstructure distributions within the electrodes.

Electrochemical reactions defined in a finite region close to the interface were implemented as various source terms in the relevant governing transport equations by [35]. The uniqueness of that approach at the macro-scale level lays in the implementation of the tortuosity factors and the material specific volume fractions for ion and electron transport within the electrodes, combined with a complete set of simultaneously solved transport equations accounting for the global internal reforming reaction kinetics. The results revealed that the electrochemical reactions only occurred within a few micro meters from the electrode-electrolyte interface. It was also found that the electrochemical reaction zone was only 2.4 μm deep in the active region adjacent to the interface in the cathode and 6.2 μm in the anode [35].

Another step in the development of numerical models of SOFC through incorporation of micro-characteristics of electrodes into a macro-model was carried out by [36]. The model treats an electrode as a reaction zone layer having triple phase boundaries. It was found that the model predicts not only the distributions of various fields within the anode, but also predicts all forms of overpotentials within the anode, determining the contribution of anode to the overall cell potential loss. For instance, it was noticed that the electronic and ionic current densities varied within the dimensionless anode thickness of 0.2, which means that for an anode thickness of 100 μm, the electrochemical reaction zone was most effective within a distance of 20 μm from the electrolyte. In addition, it was found that there were always two regions at any given conditions considered, which clearly reflects the concept of treating an SOFC electrode as two finite layers. Further, the results showed that increasing the operating temperature increases the thickness of the reaction zone layer [36].

An innovative technique for accelerating finite volume treatment of electrodes as two distinct layers, a diffusion layer and a catalyst layer, was applied in the method of macro/microstructural parameters by [37]. Electrode catalyst layers were divided into a number of control volumes arranged in a single layer grid. Variations of ADL (Anode Diffusion Layer) microstructures had different intensity effects on multi-component gas transport and the reactions involved in the process. An optimal set of the ADL microstructure was defined at the reforming reaction rate and electrochemical reactions of 0.4 and 0.2, respectively. In addition, pore size variations had significant effects on the components' diffusional resistance, reaction rates and temperature drop [37].

A further development of the microstructural approach can also be found in the work of [21], who obtained a qualitative agreement of simulated and experimental impedance spectra at different times. The simulations were carried out using the in-house software package DENIS and CANTERA software developed by [38] with the latter used to evaluate chemical source terms. The reaction mechanism consisted of 42 surface reactions among 7 gas phase species and 14 surface adsorbed species. The charge-transfer reaction mechanism comprised hydrogen oxidation at TPB by oxygen anions from electrolyte yielding water at the anode side. An elementary kinetic model was established to represent the coupled behaviour of electrochemistry, transport and degradation processes in the porous Ni/YSZ anode of SOFCs. The kinetics of carbon deposited on Ni surface was described by the following equation (2.8):

$$C_{Ni} \Leftrightarrow C_C^X + (Ni) \tag{2.8}$$

where: (Ni) denotes a free Ni surface site, C_{Ni} is adsorbed carbon species on Ni surface, C_C^X is bulk carbon, which belongs to a newly formed phase on top of the Ni surface.

All model equations were represented by partial differential equations (PDE). Spatial derivatives of the PDE system were discretised by the finite volume method. An extrapolation method based upon semi-implicit Euler discretisation was used to integrate the discretised equation system in time. Experimental electrochemical impedance spectra were simulated using a potential step and current relaxation technique developed by [39]. The impedance was obtained in the frequency domain by the Fourier transform of the resulting time-domain traces of current and potential. It was found that at OCV and a high temperature of 1023 K a surface carbon layer was

20

formed covering the Ni surface and the Ni three phase boundary and blocking heterogeneous and charge-transfer reactions. However, at a lower temperature of 923 K carbon growth proceeded inside the anode porous phase leading to significant diffusion polarisation.

In the last few years the modelling of the electrode microstructure has also been developed by means of particle-based packing algorithms. This technique was adopted for SOFCs by [40 – 43]. The microstructural simulation enabled to improve the estimation of effective properties with a reduced number of adjustable parameters. The proposed approach allows specific features of SOFC electrodes to be taken into account, such as sintering effects and particle size distributions [41]. In addition, the microstructural modelling provides a valuable tool for predicting the performance of conventional and alternative architectures [42]. Bertei et al. [44] used a Monte Carlo random walk method to calculate effective transport properties in both gas and solid phases. The electrode microstructure was modelled as a random packing of spherical particles using a sedimentation algorithm also known as drop-and-roll. More details can be found in the previous work of [45].

Model results showed that porosity lower than 0.2 and particle size smaller than 0.2 μm should be avoided in the cathode design. Smaller porosities could not guarantee a significant increase in performance and may even lead to a higher cathode overpotential [44]. In addition, Knudsen effects may arise, leading to detrimental mass transfer limitations. Furthermore, if a functional layer was inserted near the electrolyte interface, its thickness was proposed to be smaller than 2 μm since cathode overpotential sharply increases for thicknesses higher than the optimal condition [44].

2.2. Macro-modelling level

The first challenge for SOFC modelling at the macro-modelling level is the difficulty in the integration of electrochemistry, chemistry, mass transport and thermal transfer. Understanding these interactions is essential for optimising fuel cell/stack designs and their performance. In general, all SOFC processes may be summarised schematically in a block diagram presented in Figure 2.2.

At the single cell level, inputs include information on cell dimensions, cell components, physical properties of reactants and flow compositions, inlet and outlet temperatures of fuel, air and current density. At the stack level, inputs are fuel/air flow rates and fuel cell configuration in the stack. These parameters are defined in models as their boundary conditions. Inlet and outlet pressures determine fuel and air flow rates in anode and cathode channels. The flow rates significantly affect the heat exchange process inside the stack and cell temperature. Reactants pass through a porous electrode to reach reaction sites. Chemical and electrochemical processes are controlled by concentration gradients between triple phase boundaries and gases flowing in electrode channels. The external load determines the current that a SOFC produces and the rate of consumption for reactants in electrochemical reactions [3]. In general, transport of reactants from the inlet through gas channels and porous electrodes to reaction sites has to be modelled, because the electrochemical potential and reaction rates depend on the concentration of reactants. At the single cell level, electrical power and voltage are output parameters. At both levels of the single fuel cell and stack, component mole fractions and temperature distributions are also output parameters.

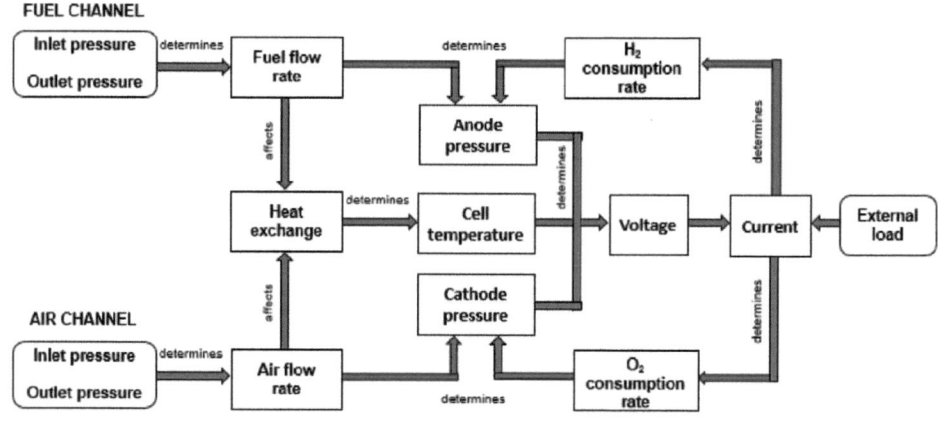

Figure 2.2. Scheme of logic correlations of SOFC process variables [3]

Several Computational Fluid Dynamics (CFD) simulation codes, such as ANSYS-Fluent, COMSOL, CFD-RC and Star-CD [14, 37, 46-50] or OpenFOAM software [51-52] have been widely used to evaluate fuel cell performance at different oxygen utilisation rates, fuel to oxygen ratios and different fuel compositions. In 2010

Andersson et al. [14] performed a review of the state-of-the-art in SOFC modelling at the macro-scale level. The authors described the mechanism behind the transport processes within SOFCs in terms of momentum, mass, heat and charge transport. In addition, approaches to model electrochemical as well as internal reforming reactions were discussed in detail. Furthermore, Hajimolana et al. [53] presented an extensive review of studies on mathematical modelling of SOFCs with respect to tubular and planar configurations divided into five subsystems and considered polarization losses, mass, energy, momentum conservations, diffusion through porous media, electrochemical phenomena in the PEN region and shift/reforming reactions inside SOFCs. A deeper insight into the SOFC modelling including the multiphysics nature of SOFCs can be found in several literature sources [53-56].

In this section, an example of the mathematical model that describes the operation of a Solid Oxide Fuel Cell is presented [49, 52]. The velocity and pressure fields in the domain are governed by the mass conservation (Eq. (2.9)) and the momentum balance of the i-th component (Eq. (2.10)):

$$\frac{\partial \varepsilon \rho}{\partial t} + \frac{\partial \rho u_j}{\partial x_j} = s_\Sigma \tag{2.9}$$

$$\frac{\partial \rho u_i}{\partial t} + \frac{\partial \left(\rho u_j u_i - \tau_{ij}\right)}{\partial x_j} = -\frac{\partial p}{\partial x_i} + s_i \tag{2.10}$$

using the following notation: ε - the porosity, ρ - the gas mixture density, u_i - the velocity component in the x_i direction x_i, p – the pressure, s_Σ - the mass source, s_i – the i-th momentum source component, τ_{ij} - the stress tensor component. The mass source, s_Σ, is the total mass consumption/production of all chemical species in a unit volume, which is non-zero in the catalyst layers due to electrochemical reactions. The porosity is usually assigned a value of unity in the gas channels, a value between 0 and 1 in the electrodes and a value of approximately 0 in the electrolyte. The stress tensor components for a Newtonian fluid were given by equation (2.11):

$$\tau_{ij} = \mu\left(\frac{\partial u_i}{\partial x_j} + \frac{\partial u_j}{\partial x_i}\right) - \frac{2}{3}\rho\frac{\partial u_k}{\partial x_k}\delta_{ij} \tag{2.11}$$

where: μ - denotes molecular dynamic viscosity of the gas mixture and δ_{ij} - the Kronecker delta. In the porous parts of the electrodes-electrolyte structure, the gas permeation is defined by Darcy's law (Eq. (2.12)):

$$\frac{\partial p}{\partial x_i} = -K_i U_i \qquad (2.12)$$

where: U_i – the superficial velocity of direction x_i, K_i – the permeability defined as (Eq. (2.13)):

$$K_i = \alpha_i |u| + \beta_i \qquad (2.13)$$

where: α_i and β_i - the coefficients assumed to be the same in all directions and defined by equations (2.14-2.15) given by [57]:

$$\alpha = \frac{1.75 \cdot \rho \cdot (1-\varepsilon)}{\varepsilon^3 \cdot d} \qquad (2.14)$$

$$\beta = \frac{150 \cdot \mu \cdot (1-\varepsilon)^2}{\varepsilon^3 \cdot d^2} \qquad (2.15)$$

where: d – the particle diameter.
The mass transport of the chemical species m in the system was represented by (Eq. (2.16)):

$$\frac{\partial \varepsilon \rho Y_m}{\partial t} + \frac{\partial \left(\rho u_j Y_m + F_{m,j} \right)}{\partial x_j} = s_m \qquad (2.16)$$

where: Y_m – the mass fraction, s_m – the mass production/consumption rate due to chemical and/or electrochemical reactions, $F_{m,j}$ – the diffusive mass flux of species m. The mass flux is modelled using Fick's law:

$$F_{m,j} = -\rho D_{m,j} \frac{\partial Y_m}{\partial x_j} \tag{2.17}$$

where: $D_{m,j}$ – the diffusion coefficient of species m, modelled often as [49]:

$$\frac{1}{D_m} = \frac{\tau}{\varepsilon} \left(\frac{1}{D_{mn}} + \frac{1}{D_{mK}} \right) \tag{2.18}$$

where: D_{mn} – the binary diffusion coefficient of species m in species n, D_{mK} – the Knudsen diffusion coefficient of species m, τ - the tortuosity. Thus, the two diffusion coefficients can be determined by the following equations (2.19) and (2.20) [46, 49, 57]:

$$D_{mn} = \frac{3}{16} \sqrt{\frac{2(R \cdot T)^3}{\pi} \left(\frac{1}{M_m} + \frac{1}{M_n} \right)} \cdot \frac{1}{N_A \cdot p \cdot \sigma_{mn}^2 \cdot \Omega_{mn}} \tag{2.19}$$

$$D_{mK} = \frac{d_p}{3} \sqrt{\frac{8 \cdot R \cdot T}{\pi \cdot M_m}} \tag{2.20}$$

where: d_p – the pore diameter, p – the total pressure, σ_{mn} - the collision diameter Lennard-Jones potential model [52], Ω_{mn} - the collision dimensionless Lennard-Jones integral [52], T – the temperature, R – the universal gas constant, N_A – Avogadro's number, M_m and M_n are the molecular weights of species m and n, respectively.

Finally, temperature distribution in a SOFC strongly influences cell performance. Accurate prediction of temperature distribution within SOFCs is essential for predicting and optimising the overall cell performance as well as avoiding thermo-mechanical degradation [58]. Most of the heat within SOFCs is generated near the electrode/electrolyte interface and is dissipated by: conduction in the solid matrix, heat transfer from the solid to the gas phase by convection within pores and gas advection through micro-pores to the flow channel. The temperature field can be resolved using the enthalpy equation (Eq. (2.21)):

$$\frac{\partial(\varepsilon\rho e)}{\partial t} + \frac{\partial(\rho u_j e + F_{h,j})}{\partial x_j} = \tau_{ij} \frac{\partial u_i}{\partial x_j} + s_h + s_e + s_c \tag{2.21}$$

where: s_h – the energy source due to ohmic resistance and radiation, s_c – the chemical energy source due to endo- or exothermic chemical processes, s_e – the heat source due to electrochemical processes, e – the specific internal energy described by equation (2.22):

$$e = c \cdot T - c_0 \cdot T_0 \qquad (2.22)$$

where: c, c_0 – constant volume specific heat [J/kg]. Thermal radiation from the surfaces of MEA (Membrane-Electrode-Assembly) is weak, therefore most authors assumed that radiative heat transfer has a negligible effect on the temperature field [59-61]. Low flow rates also cause a weak convective interaction between gas compartments and the MEA. Therefore, some authors [46] used a simple radiation model, in which the surfaces of the MEA radiate to a black-body enclosure at the nominal furnace temperature and the radiative heat flux, q_{rad}, from the MEA surfaces can be evaluated as follows (Eq. (2.23)):

$$q_{rad} = \xi \sigma_{SB} \left(T_{surf}^4 - T_\infty^4 \right) \qquad (2.23)$$

where: σ_{SB} - the Stefan-Boltzmann constant, T_{surf} - the MEA surface temperature, T_∞ - the furnace temperature, ξ - the MEA surface emissivity.

The energy source term, s_h, may be used to take into account heat generated by resistances to the flow of ions and/or electrons in the solid parts of a cell, which is called Joule heating or ohmic heat loss and can be determined from equation (2.24):

$$s_h = \frac{i^2}{\sigma} \qquad (2.24)$$

where: i – the local electronic/ionic current density, σ - the electronic/ionic conductivity. Zhang et al. [62] found that the ohmic heat loss is about 2.3 – 4.1% and 8.0% of the total heat released in a planar electrolyte-supported cell and a tubular cathode-supported cell, respectively.

The thermal flux, $F_{h,j}$, comprises Fourier conduction and heat transported by diffusive mass fluxes described as follows [49]:

$$F_{h,j} = -k\frac{\partial T}{\partial x_j} + \sum_m h_{t,m} F_{m,j} \tag{2.25}$$

where: k – the thermal conductivity of a gas mixture or of solid materials, $h_{t,m}$ – the specific enthalpy of species m.

The heat released by electrode reactions can be computed as [55]:

$$s_e = -R_e(\Delta H_e + 2 \cdot F \cdot V_{cell}) \tag{2.26}$$

where: R_e – the cell reaction rate based on the rate of change of H_2 molar concentration, ΔH_e - the reaction enthalpy change, F – the Faraday's number, V_{cell} – the operating cell voltage. The output voltage of a cell can be defined as the electric potential difference between two interconnects [52]:

$$V_{cell} = \Phi_{e,int-c} - \Phi_{e,int-a} = E_{Nernst} - \eta_{conc} - \eta_{act} - \eta_{Ohm} \tag{2.27}$$

where: $\Phi_{e,int-c}$ and $\Phi_{e,int-a}$ are the electric potential of the outermost layer of the cathode and anode interconnects, respectively.
Electrochemical reactions that take place in a hydrogen fed SOFC consist of the oxidation of hydrogen in the anode and the reduction of oxygen in the cathode, releasing a water molecule as a product in the anode:

Anode side: $\qquad\qquad\qquad H_2 + O^{2-} \rightarrow H_2O + 2e^- \qquad\qquad$ (2.28)

Cathode side: $\qquad\qquad\qquad \frac{1}{2}O_2 + 2e^- \rightarrow O^{2-} \qquad\qquad$ (2.29)

Overall reaction: $\qquad\qquad\qquad H_2 + \frac{1}{2}O_2 \rightarrow H_2O \qquad\qquad$ (2.30)

For a H_2 fed SOFC, combining the two reversible potentials gives the Nernst potential of the cell, E_{Nernst}, at the open-circuit conditions as follows [55]:

$$E_{Nernst} = E_0 + \frac{R \cdot T}{2 \cdot F} \ln\left(\frac{p_{H_2a}}{p_{H_2O,a}}\right) + \frac{R \cdot T}{4 \cdot F} \ln\left(p_{O_2c}\right) \tag{2.31}$$

where: p_{H_2a}, $p_{H_2O,a}$, p_{O_2a} - the partial pressure of the species H_2, H_2O, O_2 at the electrode-electrolyte interface, E_0 – the standard cell potential under standard conditions with the standard Gibbs free energy change of the cell reaction, ΔG^0, defined by equations (2.32) and (2.33), respectively:

$$E_0 = \frac{-\Delta G^0}{2 \cdot F} \tag{2.32}$$

$$\Delta G^0 = \mu_{H_2O}^0 - \mu_{H_2}^0 - \frac{1}{2}\mu_{O_2}^0 \tag{2.33}$$

where: μ - the chemical potential, given by $\mu = \mu^0 + R \cdot T \cdot \ln p$ for ideal gas. Concentration polarisation, η_{conc}, is due to mass transport resistance in the electrodes and can be calculated as follows [63]:

$$\eta_{conc} = \eta_{conc,a} + \eta_{conc,c} = \frac{R \cdot T}{n \cdot F} \ln\left(\frac{p^*_{H_2} \cdot p_{H_2O}^0}{p^*_{H_2O} \cdot p_{H_2}^0}\right) + \frac{R \cdot T}{n \cdot F} \ln\left(\frac{\sqrt{p^*_{O_2}}}{\sqrt{p_{O_2}^0}}\right) \tag{2.34}$$

where: p_i^* denotes the partial pressure of the i-th reactant and product at reaction sites, while p_i^0 indicates the partial pressure of the feed values of reactants and products.

Activation polarisations (activation losses), η_{act}, are the results of the kinetics involved in electrochemical reactions. The polarisations become an important loss, when the current is low, because at a low current density reactants have to overcome an energy barrier named activation energy to drive electrochemical reactions at electrodes-electrolyte interfaces and this barrier leads to the polarisation [53]. The Butler-Volmer equation is used to describe the relation between the activation polarisation overpotential, η_{act}, and the transfer current, I:

$$I = I_0 \left\{ \exp\left(\beta \frac{n \cdot F \cdot \eta_{act}}{R \cdot F} \right) - \exp\left[-(1 - \beta) \frac{n \cdot F \cdot \eta_{act}}{R \cdot F} \right] \right\} \tag{2.35}$$

where: β – the symmetry coefficient, n – the number of participating electrons involved in electrochemical processes, F – the Faraday's constant, I_0 – the exchange current density, the calculation of which for both electrodes uses the following experimental correlations [64]:

$$I_{o,a} = \gamma_a \cdot \left(\frac{p_{H_2}}{p_{ref}} \right) \cdot \left(\frac{p_{H_2O}}{p_{ref}} \right)^{-0.5} \cdot \exp\left(\frac{-E_{act,a}}{R \cdot T} \right) \tag{2.36}$$

$$I_{o,c} = \gamma_c \cdot \left(\frac{p_{O_2}}{p_{ref}} \right)^{0.25} \cdot \exp\left(\frac{-E_{act,c}}{R \cdot T} \right) \tag{2.37}$$

where: γ_a and γ_c are the anodic and cathodic pre-exponential coefficients, $E_{act,a}$ and $E_{act,c}$ – are the anodic and cathodic activation energies.

The Ohmic polarisation is mainly caused by the resistance to the transport of oxygen anions through the electrolyte, η_{Ohm} and can be estimated as:

$$\eta_{Ohm} = \frac{\delta \cdot l_e}{\sigma_e} \cdot I_e \tag{2.38}$$

where: l_e – the electrolyte thickness, σ_e - the electrolyte anionic conductivity, I_e – the mean current density in the electrolyte estimated as follows:

$$I_e = \frac{I_a + I_c}{2} \tag{2.39}$$

To transport oxide ions from the electrolyte to the anode (Eq. (2.29)) and to block electron flow from the anode to the cathode are the main functions of a SOFC electrolyte. The flow of electronic charges through external circuit balances the flow of the ionic charge through the electrolyte and thus the electrical power is produced.

This transport can be described by considering the ionic and electronic charge transport equations (2.40-2.41):

$$\nabla \cdot i_{ion} = Q_{ion} \qquad (2.40)$$

$$\nabla \cdot i_{elec} = Q_{elec} \qquad (2.41)$$

and the Ohm's law for ion and electron transfer in an electrolyte (Eq. (2.42-2.43)):

$$i_{ion} = -\sigma_{ion} \nabla \phi_{ion} \qquad (2.42)$$

$$i_{elec} = -\sigma_{elec} \nabla \phi_{elec} \qquad (2.43)$$

where: ϕ - the ionic/electron potential, σ_{ion} - the ion conductivity in the electrolyte material, σ_{elec} - the electronic conductivity in the anode, $\sigma_{elec,a}$, or in the cathode, $\sigma_{elec,c}$. The interfaces/boundaries can be defined as electric potential ($V=V_0$), electric isolation ($n \cdot i = 0$) or ground conditions ($V=0$) [14]. The effects of electrochemical reactions are either defined as source terms in governing equations or as interface conditions at the electrolyte/electrode interface as was discussed by [65]. The presented model equations describe the relationship between the cell current and voltage.

The whole model presented in this section allows to calculate the 3D distribution of gas-phase species, current, potential as well as pressure and flow velocities for SOFCs. A detailed model, which takes the form of a partial differential equation system, is stiff and therefore efficient and specialised numerical techniques are needed to solve the equations. In addition, various sub-models are coupled through appropriate source terms, which means that transport properties influence species distributions and consequently influence charge-transfer kinetics, which again influences concentration gradients. Interrelations between the discussed transport equations and some calculated parameters of a SOFC are presented in the first two blocks on the left in Figure 2.3.

In the recent decade several other modelling alternatives for SOFCs including physical models were developed. Wang et al. [66] defined three main categories of modelling: grey, black and white-box models. The grey-box method based on a combination of a priori knowledge concerning the process and mathematical relations, which describe the behaviour of a SOFC system. Black-box models predicted fuel cell performance without prior knowledge of many physical, chemical and electrochemical parameters. Black-box models are usually based on artificial intelligence such as Artificial Neutral Networks or neuro-fuzzy systems [67-68]. White box models were of two types: physical models and equivalent circuit models based on Electrochemical Impedance Spectroscopy (EIS) technique [69].

The second major challenge at the macro-modelling level is optimisation of fuel cell durability and performance stability of a fuel cell during its operation. Performance instability of a SOFC is due to mechanical instability of its structure subjected to moderate stress. Stress applied to ceramic components can arise from manufacturing (residual stresses), differential TEC (Thermal Expansion Coefficients) of cell layers, spatial or temporal temperature gradients, oxygen activity gradients and external mechanical loading [70].

Predictions of thermal stresses in Solid Oxide Fuel Cells are based on thermal-fluid models predicting fuel cell operation, while mechanical properties of the anode and cathode are determined theoretically through composite structure approximation. An example of coupling a flow solver, a SOFC module and a stress solver is presented in Figure 2.3 [71].

Nakajo et al. [72] used temperature profiles generated by a thermo-electro-chemical model implemented in gPROMS to calculate thermal stress distribution in a tubular SOFC. Solid heat balances were calculated separately for each layer of the Membrane-Electrode-Assembly (MEA) in order to detect radial thermal gradients. The stress field model was based on the static uncoupled linear thermo-elastic theory. Stress fields showed high tensile stresses in limited areas at the ends of the cell in the electrolyte and significant values in the whole anode. The latter effect occurred as a result of a high anode coefficient of thermal expansion. The magnitude of stresses was higher at the fuel/air inlet, where the internal steam reforming reaction induces a temperature drop.

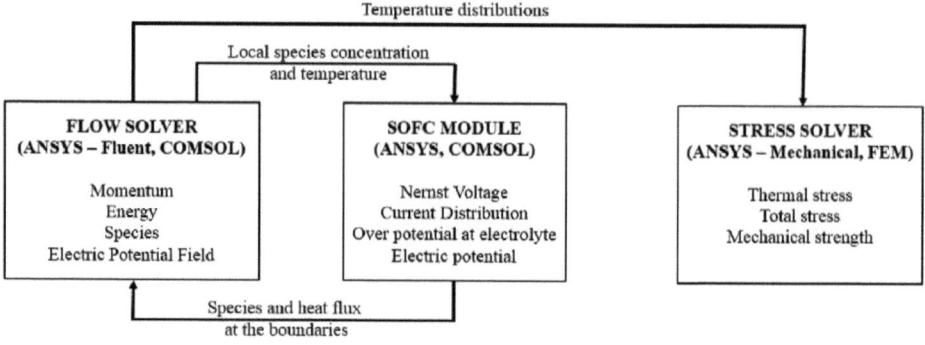

Figure 2.3. Coupling of a flow solver, a SOFC module and a stress solver [71]

Chiang et al. [73] employed the commercial CFD code Star-CD with the es-sofc module to simulate current-voltage characteristics and to provide temperature fields of the cell to the commercial code MARC for thermal stress analysis. To resolve the incompatibility issue between the Star-CD and MARC software products a pre-processing software PATRAN was used. The relationship between nodal displacement and element strain was expressed as a product of matrix related to nodal displacement and element strain, where the strains were represented in terms of stresses. A difference in experimental data and simulation results of current density was about 9%. Cell performance improved as gas flow rate increased. A lower flow rate decreased the electrochemical reaction rate. Thermal stresses were much higher at a lower operating voltage of 0.5 V. The maximum principal stress decreased as operating voltage increased. Moreover, the maximum principal stress increased as the temperature differences increased. The average strength of an anode supported Positive-Electrolyte-Negative (PEN), depending on its composition, was around 71-187 MPa [73]. It was also found that principal stresses were lower for the inlet gas temperature of 790°C. Hence, it would be beneficial to preheat inlet gases to a certain extent to minimise temperature gradients and subsequently reduce thermal stresses of cell components. Chiang et al. [73] reported that temperature difference should be controlled within 50°C on the PEN surface for the sake of structure integrity.

A capability of performing steady state and transient thermal stress analysis in a planar bipolar SOFC was also developed by Selimovic et al. [74]. The approach was based on the coupling of electrochemical-thermal FORTRAN code with the Structural

Mechanics Module in the commercial finite element tool FEMLAB. The model could only solve mechanical stress problems in the three layered structure and the impact of the interconnect layer was accounted for in the heat balance of the solid and in the ohmic resistance term of the electrochemical-thermal model. It was found that the largest stress occurred in a ceramic fuel cell fuelled with pre-reformed methane and it was located in the electrolyte layer at the interface with the anode. The stress was estimated slightly above the strength of the electrolyte material [74].

To sum up, only few research studies [48, 50, 72-75] examined thermal stresses in SOFCs, which may have resulted from the fact that there are many degrees of freedom in a design (tubular or planar, anode or cathode supported) as well as precise data of the mechanical properties of the MEA materials are rare, but also some other uncertainties remain in critical areas, such as the temperature at which no internal stresses occur.

3. Stack level modelling

Numerous simulation models have been developed to predict the effects of various stack geometrical and operating parameters of SOFC. The objectives of those numerical studies were to maximise power density and fuel utilisation and to minimise non-uniform current density and temperature distributions [70-71, 76-78].

3.1. Heat transfer

A thermal model for a SOFC stack may include the following components: (i) heat release and absorption stemming from reforming, shifting or electrochemical reactions as well as (ii) heat transfer by convection between the solid and the gas phases and (iii) radiation heat transfer between the solid part and the gas [79]. Formulation details related to the thermal model for a SOFC can be found in some publications on the subjects [53, 80-82].

One of the first studies on heat transfer in the Solid Oxide Fuel Cell modelling for both planar and tubular configurations was carried out by VanderSteen and Pharoah [83]. It demonstrated that radiation has to be considered as a non-negligible part of the SOFC issue. The authors presented a thermal transport model that included conduction, convection and radiation in the participating media using a commercial CFD code. For the planar SOFC, model results showed that temperature gradients were greatly affected by radiation. When radiation heat exchange was modelled, lower wall temperatures in the cell were noticed, which means that radiation effectively decreased temperature gradients in the cell. This was a significant result, because it clearly demonstrated that ignoring radiation heat transfer in a thermal model of a fuel cell can result in the calculated temperature being over 30 degrees too high [83]. Similar results were obtained for a tubular SOFC, where temperature gradients were greatly affected by radiation. In addition, it was found that larger values of gas absorptivities led to lower overall temperatures.

Hartvigsen et al. [84] demonstrated significant changes in stack temperature with the inclusion of radiation effects. However, authors did not provide details of the radiation model used in the analysis. Yakabe et al. [85] found that the implementation of radiation heat exchange within the flow channels resulted in a flatter distribution of the temperature profile along the fuel cell. However, it should be mentioned that their model neglected absorption, emission and scattering in the media and it took into account only radiation effects for surface-to-surface transfer.

Tanaka et al. [82] developed a new three dimensional simulation code of the planar SOFC stack to calculate accurately the effect of radiation heat transfer from cell stack surfaces. The ambient temperature of the cell stack was assumed to be constant and radiation heat transfer was calculated under the condition that the cell stack was wholly surrounded by a wall with this ambient temperature. The effect of heat convection with the gas flow and heat transfer between the solid and gas were taken into account by including them in the heat generation term of the heat conduction equation. It was found that the influence of radiation heat transfer was very large near the surfaces of the cell stack and cell temperature at the top, bottom and side surfaces was mainly determined by the ambient temperature. Therefore, the central part of the cell stack was almost free from the influence of radiation heat transfer. The authors [82] concluded that modern CFD packages are easily capable of including radiation models in SOFC analysis.

However, many authors neglected the effect of radiation heat transfer [61, 86]. Daun et al. [61] state that due to the minimal effect of thermal radiation on the temperature field in the planar SOFC, thermal radiation can be excluded from a detailed CFD analysis of a planar anode supported SOFC. The importance and nature of thermal radiation was determined by comparing temperature distributions obtained using the Schuster-Schwartzchild/conduction and Monte Carlo/conduction solvers coupled to the one calculated by assuming conduction as the only mode of heat transfer [61].

In contrast to these studies, Murthy and Fedorov [87] found that ignoring radiation effects caused errors of 100°C in the surrounding temperature fields. The Schuster-Schwartzchild two-flux approximation was used for treating thermal radiation transport in a thin YSZ electrolyte and the Rosseland radiative thermal conductivity was used to account for radiation effects in Ni-YSZ and LSM electrodes. Thermal radiation heat transfer was coupled to overall energy conservation equations through the local radiative flux. A single cell model represented a repeating unit cell in the center of a large stack. Based on simulation results from Fluent CFD software, it was concluded that radiation heat transfer effects were significant and need to be accounted for in accurate predictions of temperature field and fuel cell output voltage [87].

Damm and Fedorov [59, 60] found that the effect of radiation accounted only for a few degrees in cell temperature field and the heat was predominately removed from cells by convection. This opposite hypothesis to their previous results [87] was explained by the fact that different SOFC geometries were involved in both studies.

Kattke et al. [88] reported that radiation heat transfer was the dominating mechanism of a 66-tube SOFC stack cooled within tubular stacks. Radiation accounted for 62-93% of the total heat evacuation from the external tube surface. The dominance of radiation led to a strong relationship between the power output of a tube and its view factor from the tube to the relatively cold cylinder wall surrounding the bundle. Recuperative heat exchange between the SOFC tail-gas and inlet cathode air and reformer air/fuel preheat processes were captured within the CFD model.

Some simulation results seem to contain significant inconsistencies. Nevertheless, taking into account how difficult it is to measure local temperatures inside the stack since SOFC stacks operate at high temperatures, it seems numerical modelling of SOFC heat transfer and simulations are crucial for optimisation of geometry and system performance. It was shown in [89] that fuel and air temperature as well as the heating rate of a SOFC have a direct impact on the operating lifetime, performance of the cell stack and material properties including micro-cracking. Therefore, thermal modelling is an integral part in the development of the SOFC technology.

3.2. Thermo-mechanical models

Thermal stress analysis is a very important method to study the performance of a system operating at high temperatures typical for SOFCs working conditions. High temperature may cause damage of a fuel cell stack due to stress concentration problem. Therefore, several studies have used the Finite Element Method (FEM) to predict the thermal stress distribution to better understand details of internal processes occurring within the SOFC and for a better design of the system [70-71, 77-78].

Similarly to Chiang et al. [73], Wei et al. [71] performed simulation of thermal stress for a new design of counter-flow channels in a planar SOFC stack based on the conceptual coupling of the ANSYS-Fluent flow solver with the SOFC module and the ANSYS stress solver. The flow solver provides species and temperature distributions to the SOFC module, while the SOFC module returns species and heat fluxes at the boundaries. The obtained temperature distributions were input into the ANSYS stress solver to calculate stress distribution, which includes the von Mises stress in metal and the maximum principal stress in ceramic material. The effects of the geometry of flow channels, interconnect, inlet and outlet port designs and stress distributions on the SOFC design were numerically investigated for a planar anode supported SOFC stack. To simplify the SOFC model, the cell stack was divided into interconnect, cell and flow channel components. Porous electrodes were considered as isotropic and homogeneous and the performance of a cell stack was estimated by simulation of a single cell. Then, Wei et al. [71] simulated the thermal-fluid field of a single and three-cell stack. Finally, they simulated an H type three-cell stack to obtain stress distribution at 700, 800 and 900°C for different materials (stainless steel, Grancrete and others). Results showed that the use of Grancrete cell support effectively reduced the maximum principal stress of the cell. It was also found that both stainless steel and Grancrete cell support can withstand the thermal stress under the specified operation of the cell.

Nonetheless, the available literature shows that a detailed simulations of thermal stress and its damaging effects on SOFCs are still in an early stage of development [70, 77-78]. In most studies the values of mechanical and thermal properties of SOFC components were assumed constant. According to [77] these characteristics depend essentially on temperature and porosity. Mounir et al. [70] showed that temperature varies on the fuel cell surface, thus the characteristics of materials can not be considered as uniform. Therefore, temperature profiles generated in simulations by thermo-electrochemical model were applied to calculate thermal stress distributions in SOFCs. The model was developed only for cases of linear response for all components. All used materials had isotropic and uncoupled thermal expansion attributes. The thermo-elasticity theory was applied to address the effects of temperature, moisture and mechanical forces on the behaviour of elastic bodies. It was found that the coefficient of thermal expansion and the Young modulus mismatch between layers had a greater influence on the strain of cell components and hence on the probability of survival. The survival chance of the electrolyte was lower than that of other components.

Extended study on modelling that combines thermo-electrochemical models accounting for cell degradation and a contact thermo-mechanical model that considers rate-independent plasticity, creep of the components materials and shrinkage of the nickel based anode during thermal cycling of SOFC stacks was presented by Nakajo et al. [77]. The modelling approach was described in detail elsewhere [90]. A percolation model predicted reduction of the triple phase boundary length due to the growth of nickel particles in the anode. A semi-empirical relation was assumed for the evolution of the nickel particle radius that depended upon temperature, steam and hydrogen partial pressure due to the mechanical constraints imposed by the YSZ network. XRD stress measurements in the electrolyte of anode/electrolyte bilayers from the literature [90] were used to validate calculated residual stresses in cell layers as a function of temperature. It was noticed that a detrimental evolution of a temperature profile can induce unexpected failure if control strategy of electrical load demand is not adapted for degradation in the long term. The direct influence of creep deformation on the integrity of the cell was significant with glass-ceramic sealants, resulting in computed factors of 0.02 up to 1.3 for the anode probability of failure, and of 4 to 136 for that of the cathode. However, it was rather limited for compressive sealing gaskets. The residual stress in the cell and creep relaxation in the metallic interconnect increase in the long-term the deflection of a SRU (Standard Repeating Unit) in a stack. The tensile stress in the high temperature zones was relieved, whereas that in the cold ones increased [90].

In the majority of the discussed papers, the level of coupling between thermo-electrochemical and mechanical aspects was restricted to discrete importing of temperature profiles, which had to be further modified to improve and address the complexity of the real situation.

4. System level modelling

The objective of the present section is to summarise features of several system level computational models with different reforming techniques and recirculation of anode and cathode gases. The section also contains comparative analysis of different system configurations and operating conditions. In the literature one can find five main types of integration component models capable to accurately predict the performance of a SOFC stack and remaining Balance of Plant (BoP) components. Each system model is based on different fuel processing methods.

A simplified scheme of a SOFC system is shown in Figure 4.1. It illustrates a typical power system consisting of a SOFC stack and the following system units: reformer to fuel processing , air preheater, off-gas burner to burn anode exit gases, heat exchanger for internal heat utilisation, and DC/AC converter.

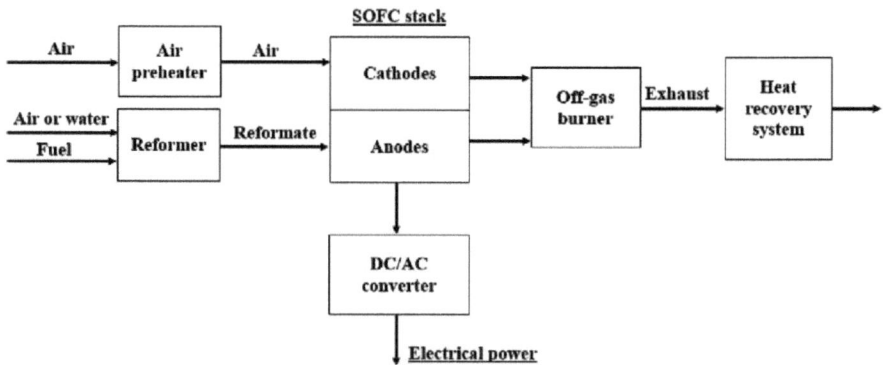

Figure 4.1. Scheme of the SOFC system

SOFC stack and afterburner release heat, while the reformer may work exothermically, endothermically or autothermically. The most important reforming methods for fuel cell applications are: Catalytic Partial Oxidation (CPO$_x$), Steam Reforming (SR) and Authothermal Reforming (ATR), which are shortly discussed below.

4.1. Catalytic Partial Oxidation (CPO$_x$)

In the first approach Partial Oxidation Reforming (CPO$_x$) exothermic method is applied. An example of a schematic diagram of a CPO$_x$ based on SOFC system is presented in Figure 4.2.

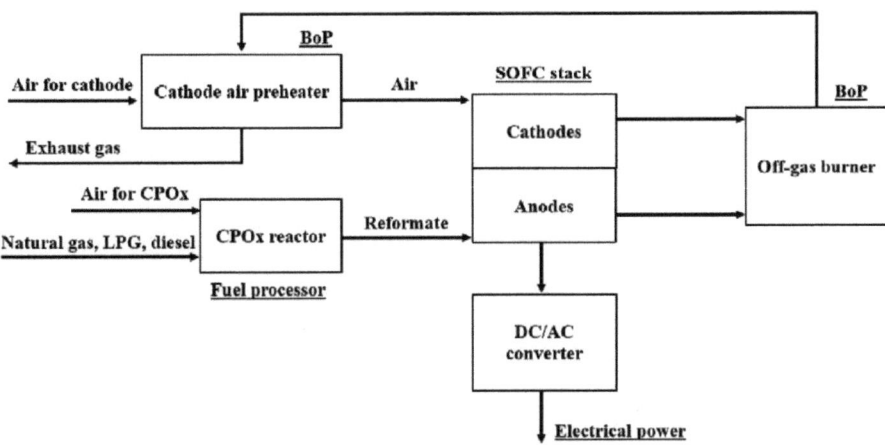

Figure 4.2. Schematic diagram of the CPO$_x$ based on SOFC system [4, 10]

Oxidation reaction can take place in the presence of a catalyst or without it. Operation of the SOFC stack with fuel utilisation factor of 0.7-0.85 enables burning the depleted gas in an off-gas burner. Heat carried by gases exiting the off-gas burner is used for air preheating in a cathode air preheater. The CPO$_x$ system can be fed using hydrocarbons such as methane, propane, butane, LPG, biogas or diesel. However, using more complex hydrocarbons is conducive to soot formation. This is the main reason why the reformate temperature should be kept in the temperature ranges of 750-800°C. According to [91] temperature control is complicated task particularly in small scale systems and requires preheating of air or fuel mixture. Therefore, one of the key parameters in the CPO$_x$ system is an air to fuel ratio, λ, defined by equation (4.1) [4]:

$$\lambda = \frac{x_{O_2} \dot{V}_{air}}{2\dot{V}_{CH_4} + 3.5\dot{V}_{C_2H_6} + 5\dot{V}_{C_3H_8} + 6.5\dot{V}_{C_4H_{10}}} \qquad (4.1)$$

where: $\dot{V}_i$ - volumetric flow rate of the i-component, x_{O_2} - volume fraction of oxygen. The air to fuel volume ratio should be in a narrow operation range of 0.27-0.34. The lambda coefficient, $\lambda > 0,3$ gives a high stack temperature of around 950°C, which requires accurate flow measurement. To obtain correct gas compositions, a sensitive λ control is necessary for the CPOx reformer reactions (4.2):

$$C_xH_y + \frac{x}{2}O_2 \leftrightarrow xCO + \frac{y}{2}H_2 \qquad \Delta_r H^0 < 0, \text{ exothermal} \qquad (4.2)$$

The main advantages of the CPO_x system are reduced number of components and low investment costs. The main disadvantage is limited electrical efficiency of 35% in appropriate power ranges. Concerning the practical development of PO_x reactors, the major challenges result from slow kinetics and the decrease of flame stability at low adiabatic flame temperatures. However, a proposal to overcome barriers was presented by Al-Hamamre et al. [92]. Taking into account that a higher reaction rate should be expected in porous media than in normal combustion processes, Al-Hamamre et al. [92] studied numerically and experimentally the partial oxidation process of methane in an inert porous reactor. The process was performed with preheated air up to 700°C premixed with methane before entering the porous reaction zone. Different highly porous structures were used: Al_2O_3 fiber static mixer structures and SiC foams. Thermodynamic calculations and kinetic simulation were performed using CHEMKIN software. Numerical results showed that the air ratio down to 0.4 was a proper limit to perform the PO_x process, which is consistent with the previous authors [4, 91]. For lower air ratio values, higher preheating temperature was needed to achieve a higher reaction temperature. Higher heat recuperation was detected in the case of SiC foam based reformer than in the case of Al_2O_3 fiber static mixer structures. Regarding the soot point, the SiC foam based reactor showed better performance and the soot point was 0.42, while for Al_2O_3 mixer structures it was 0.45 [92]. Similar studies on partial oxidation of methane in an adiabatic type catalytic reactor were performed by [93]. The authors examined four catalysts in CPO_x processes in the range of various ratio of methane/air from 2.2 to 6.0. Measurements and modelling agreed at temperature close to that measured at the catalyst outlet.

Detailed characteristics of all major system components of a SOFC based micro-Combined Heat and Power (CHP) system for natural gas with CPO_x as reforming technology were presented by [10] in the framework of the European project FC-DISTRICT. Heat and mass balance for the overall system was generated using Aspen simulation. Simulations were carried out at a nominal stack power of 1.5 kW_{el} with turndown ratio of 3:1. The air/fuel ratio in the CPO_x reforming process was around 0.31. Different natural gas types such as methane, propane and synthetic biogas were used as a fuel. The electrical efficiency for stack power of 1.5 kW_{el} was calculated as approximately 32%, while the thermal efficiency for combined hot water storage and floor heating application at 1.5 kW_{el} was around 55%. The net system efficiency was in the range of 83-87% for 1.5 kW_{el} stack power for examined fuel types [10].

An interesting research study was presented by Wang and Cao [94] carried out for the first time for partial oxidation of butanol in a SOFC system. The model was based on Gibbs free energy minimisation thermodynamic method. The partial oxidation process of butanol was lightly exothermic ($\Delta H_{298}^{o} = -115.80$ kJ/mol) and produced H_2 and CO. In order to maximise H_2 production it was confirmed that enough O_2 had to be supplied to minimise butanol dehydrogenation, dehydration and decomposition. Almost complete conversion, 93.07% of H_2 and 94.02% of CO was obtained at the optimised conditions: temperature of 1115 K, pressure of 1 atm and $O_2/C_4H_{20}O$ molar ratio of 1.6. It is surprising that under the optimised conditions the authors [94] reported that energy efficiency was 88.78%. However, one should bear in mind that the thermodynamic equilibrium analysis did not consider any kinetic constraints, such as concentration and temperature gradients occurring in real processes, which can explain such a high value of the reported efficiency.

4.2. Steam Reforming (SR)

The second fuel processing technique widely used for hydrocarbons is Steam Reforming (SR) or wet reforming presented in Figure 4.3. Water used for steam reforming has to be deionised and demineralised.

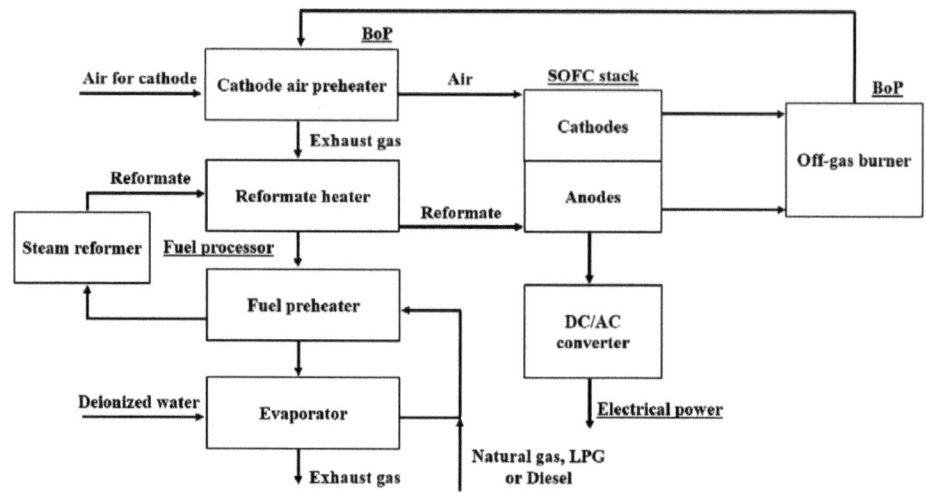

Figure 4.3. Schematic diagram of the SR based on SOFC system [95]

The main advantage of the method is electrical efficiency of approximately 45-60%. The endothermic reaction of hydrocarbons with steam generates carbon monoxide and a hydrogen rich gas. Ideal steam reforming process for hydrocarbon fuels can be described by reactions (4.3) - (4.4):

$$C_xH_y + xH_2O_{(g)} \leftrightarrow xCO + \left(\frac{y}{2} + x\right)H_2 \tag{4.3}$$

$$CO + H_2O_{(g)} \leftrightarrow CO_2 + H_2 \tag{4.4}$$

Typical steam reforming reactions for methane can be written as following (4.5) - (4.6) [96]:

$$CH_4 + H_2O \leftrightarrow CO + 3H_2 \tag{4.5}$$

$$CO + H_2O \leftrightarrow CO_2 + H_2 \tag{4.6}$$

Steam to carbon volume ratio can be described by equation (4.7):

$$S/C = \frac{\dot{V}_{H_2O}}{\dot{V}_{CH_4} + 2\dot{V}_{C_2H_6} + 3\dot{V}_{C_3H_8} + ...}$$

(4.7)

An internal reforming in a SOFC stack is often used to decrease system complexity through a reduction of air demand. However, the main problem of internal reforming is the mismatch between the heat requirement for steam reforming and the heat available from the fuel cell section. The internal reforming operation could lead to local sub-cooling around the entrance area of reformer part, which can result in mechanical failure due to thermally induced stresses [97]. An additional problem related to internal reforming is carbon deposition on the anode site, which occurs due to cracking reactions. Carbon formation could result in the deactivation of anode material, which leads to the loss of fuel cell performance [21]. The quantity of carbon deposited on the anode is affected by the operating temperature and the methane to steam ratio [98-99].

A further disadvantage of internal reforming in comparison to external reforming is that fuel cell materials may be poisoned by some impurities such as sulphur components in the feed fuel [97].

Liso et al. [96] presented performance comparison between partial oxidation, discussed in the previous paragraph, and methane steam reforming processes with recirculation of anode and cathode gases used for a micro Combined Heat and Power (CHP) system for residential application. CHP system with a SOFC stack was fuelled by natural gas. The thermochemical fuel cell sub-model based on chemical equilibrium and conservation of elements. Steam reforming and water gas shift reactions were considered in addition to the electrochemical oxidation of hydrogen. Both reactions were carried out over a supported nickel catalyst at temperatures above 500°C. The chemical reactions in both systems: POx and SR were modelled in chemical equilibrium using standard Gibbs free energy. Each configuration was optimised to obtain the maximum electrical efficiency with 1 kW electric output and 2 kW heat output. The SR configuration showed a better electrical efficiency for a higher fuel conversion efficiency in comparison to the system with POx. Using anode gas recirculation the stack fuel utilisation was reduced from 80% to 60%, at constant 80% fuel utilisation of the system. The use of anode gas recycle increased the anode inlet gas flow, lowering cell fuel utilisation at a given current [96]. Additionally, reduction in a water heating system, in heat exchanger dimensions and fuel cell dimensions reduced system capital cost. On the other hand, cathode gas recirculation reduced the fresh air feed required to cool the stack lowering the blower capacity and reducing the cathode air heat exchanger dimensions.

4.3. Steam reforming with recirculation of anode off-gas -"hot" blower

Endothermic anode off-gas recycling presented in Figure 4.4 is another fuel processing technique. It combines an easier system setup of steam reforming and a high efficiency of above 50% obtained by recirculation of anode off-gases.

Recirculation of gases exiting anode enables to overcome difficulties caused by water preparation required for conventional steam reforming [100]. Reactions occurring in a SOFC stack generate water in the form of vapour, therefore recirculation of 30-40% of the anode off-gas supply is sufficient for steam reforming. A proper oxygen to carbon ratio defined by equation (4.8) can be maintained by a wye in the system between off-gas burner and SOFC stack [91]:

$$O/C = \frac{2\dot{V}_{CO_2} + \dot{V}_{CO} + \dot{V}_{H_2O}}{\dot{V}_{CH_4} + \dot{V}_{CO_2} + \dot{V}_{CO}} \tag{4.8}$$

Estimation of the O/C ratio is a tricky task especially at high temperatures similarly to problematic measurement of components, which are main disadvantages of the anode off-gas recirculation system. Dynamic system behaviour and long-term stability of a power unit still remain significant challenges.

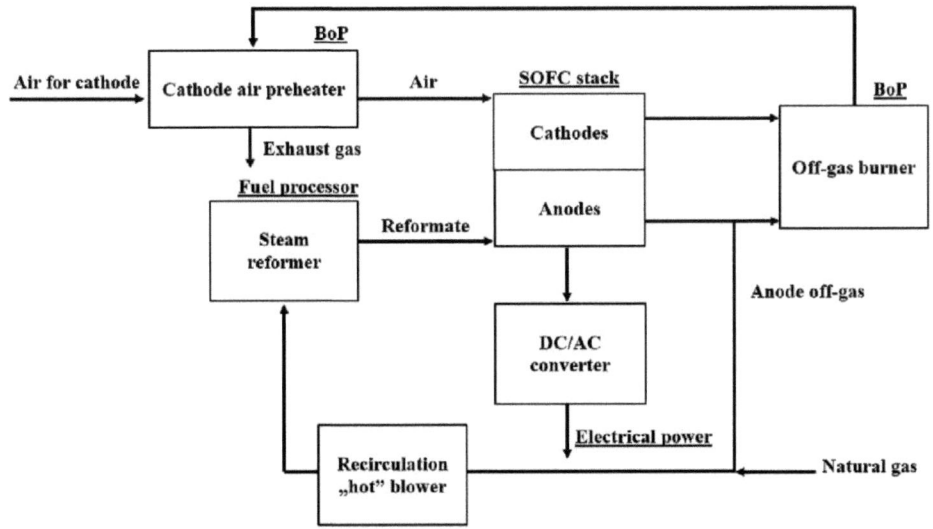

Figure 4.4. Schematic diagram of the SOFC-based system with steam reforming with recirculation of anode gases [91, 100]

A thermodynamic and electrochemical model was developed by [101] to analyse a combined system of a SOFC coupled with a fuel reforming device using the anode off-gas recycle. A temperature dependent equilibrium constant was obtained using the classical method in which the change in Gibbs free energy of the reactions was used. A direct internal reforming SOFC model was calculated and CO and methane were assumed to react with the water vapour for producing more hydrogen in the anode. The system performance was evaluated as a function of steam to carbon ratio, fuel cell temperature, anode off-gas recycle ratio and CO_2 adsorption percentage. The authors found that the system efficiency started at around 70% and then monotonically decreased to the average of 50% at the peak power density before dropping down to zero at the limiting current density point [101]. The main finding was that the optimal system was the one when the SOFC operated at around 900°C with the steam to carbon ratio higher than 3, maximum CO_2 capture, minimum anode off-gas recirculation and the system efficiency peaked.

4.4. Steam reforming with anode off gas recirculation – "cold" anode off gas blower

Control of the recycled stream can be carried out by recirculation blower operating in higher or lower temperature obtained by using an additional pre-reformate heater as shown in Figure 4.5.

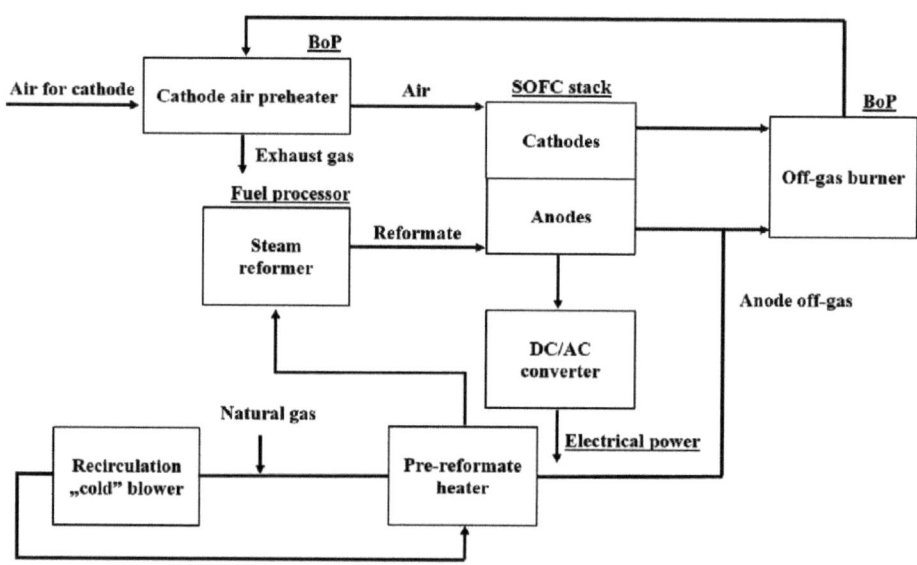

Figure 4.5. Schematic diagram of the SOFC-based system with steam reforming with a "cold" anode off gas blower [91, 102]

An additional heat exchanger allows to reduce temperature in the system if reduction of gas temperature in the blower is needed. The anode off-gas is mixed with fuel and is supplied to the steam reformer by a high or low temperature fuel recirculation blower. Kupecki et al. [100] analysed the technology in a SOFC-based micro-scale CHP (Combined Heat and Power) system with electrical power output of 0.7 – 3 kW fed by biogas. A numerical model of CHP unit was defined in ASPEN Plus modelling environment. The authors [100] achieved electrical efficiency of 38.8% with overall system efficiency of 67.8%.

4.5. Steam reforming with anode off gas recirculation with ejector

An alternative to the SOFC system based on steam reforming with a low temperature blower for recirculation of anode off-gas discussed above is steam reforming with ejector based recirculation of anode gases. The main advantage of the system is a very simple system layout, reduction of complexity and elimination of a recirculation blower. A schematic diagram of the system with ejector for recirculation of anode off-gas is presented in Figure 4.6. Disadvantages of the SOFC based system with the steam reforming and an ejector supporting recirculation are limited modulation ratio and high gas inlet pressure required. However, numerical investigations carried out by [8] showed that it is desirable to operate the system at the lowest feasible S/C ratio, compatible with the problem of carbon deposition to be able to reduce the fresh fuel pressure. Therefore, according to [103] the S/C ratio should remain at low values as long as it can avoid carbon formation and thermal gradient limits of the SOFC.

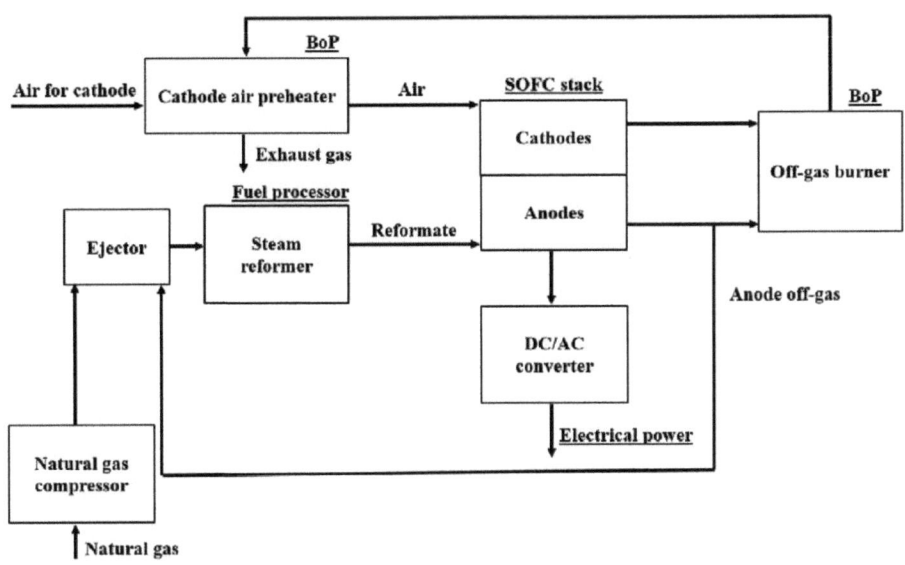

Figure 4.6. Schematic diagram of the SOFC-based system with steam reforming and an ejector base recirculation [91]

4.6. Autothermal reforming (ATR)

Autothermal reforming (ATR) is another available processing technology, which enables utilisation of diesel fuel. A schematic diagram of an autothermal reforming system is presented in Figure 4.7.

The mechanism of autothermal reforming can be represented by the following chemical reaction scheme (4.9) [104]:

$$C_xH_yO_p + Steam + air \rightarrow H_2 + carbon \ \ oxides + N_2 \qquad (4.9)$$

Autothermal reforming system was well characterised by [105], where it was used to reform desulfurised JP-8 fuel to provide H_2 and CO as fuel to planar and tubular Solid Oxide Fuel Cell stacks.

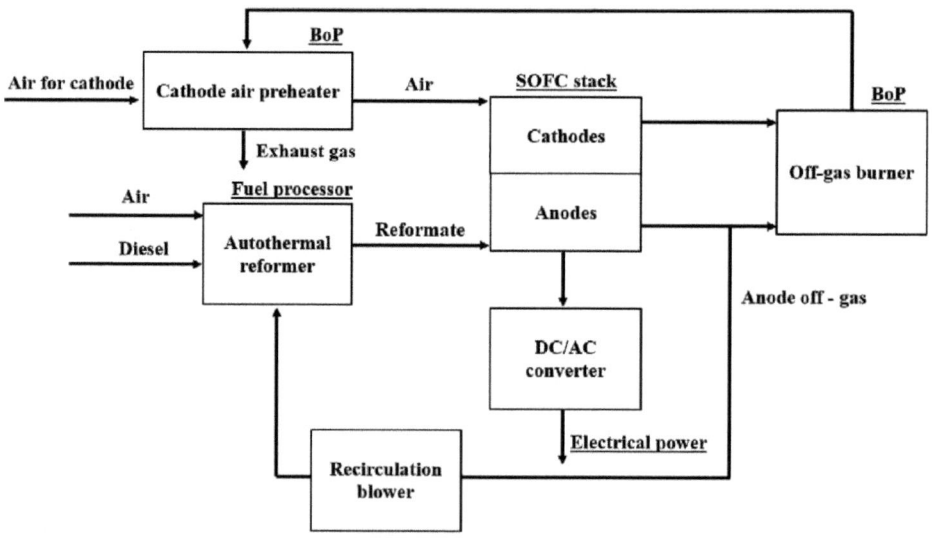

Figure 4.7. Schematic diagram of the SOFC system based on autothermal reforming [91]

It was found that increasing S/C ratio and ATR operating temperature helped suppress coke formation, but at the same time increased fuel dilution. This caused decay of Nernst voltage and efficiency. The maximum and minimum total efficiency of the system was 70.2% and 61.7% at the 700°C and 800°C ATR operating temperatures at S/C ratio of 0.4 and 1.0, respectively [105].

Ersoz et al. [106] investigated a SOFC micro cogeneration system for power generation using Aspen-HYSYS 3.2 simulation software. Natural gas was simulated in three different fuel reforming technologies for hydrogen production such as: autothermal reforming, partial oxidation and steam reforming. The simulated SOFC system consisted of three sections: fuel processing section, fuel cell section and auxiliary units. Thermodynamic equilibrium system calculations were based on minimising the Gibbs free energy. During simulations, equilibrium temperature and outlet compositions of each reactor were calculated. The investigated ranges of operating conditions for the autothermal reforming were as follows: S/C of 0.3 – 3, O/C of 0.35 – 0.6 [106]. Obtained simulation results indicated that the fuel processing efficiencies decreased in the order: steam reforming > autothermal reforming > partial oxidation. The overall system efficiency of natural gas fuel was the highest for the steam reforming (41%), a little bit lower for the autothermal reforming (38%) and the lowest value of 29% was obtained for the CPOx. The optimum S/C ratio at 3.5 was obtained to fulfil the requirements for temperatures around 800°C for steam reforming process. The optimum O/C and S/C ratios were 0.45 and 1.5, respectively, for autothermal reforming at the inlet temperature of 700°C.

A comparison of SOFC systems with different fuel processing methods is presented in Table 4.1.

Combined heat and power systems using high temperature SOFCs were simulated with commercial process modelling software, such as Dynamic Network Analysis (DNA) [107], ECLIPSE [108] or Aspen Plus™ [8, 12, 103, 109-110], HYSYS [111-113].

Table 4.1. Comparison of SOFC systems with different fuel processing methods [91]

	Catalytic partial oxidation (CPOx)	Steam reforming (SR)	Anode off-gas recirculation with steam reforming
Electrical efficiency	30-35%	45-60%	50-60%
Overall efficiency	85-95%	85-95%	85-95%
Power to heat ratio	0.5-0.7	1.0-2.2	1.3-2.3
Heat requirement	exothermic	endothermic	endothermic
Air/fuel ratio	0.27-0.31	-	-
Sensitive to gas and water quality	no	yes	partly
Start-up	fast	external igniter need	external igniter need
Transient conditions	easy to control	relatively stable	relatively stable
System costs	low	high	high
Number of components	minimal	large	large
Dedicated applications	μCHP system 0.5-5 kW	residential ➢ 1.5 kW	large system ➢ 50 kW

The DNA is a simulation tool developed in the Thermal Energy Section of the Technical University of Denmark [107, 114]. The DNA was based on a model formulated by connecting components with nodes and adding operating conditions to build up the system. The program is written in Fortran and its users add additional components and thermodynamic state models to present libraries.

The ECLIPSE process simulation package enables to perform a full economic analysis of various systems. An example of how the ECLIPSE can be used was presented by [108]. The study considered a combined biomass gasification with SOFC stacks. From ECLIPSE simulations the efficiency of the 250 kWe system was found to be around 39%, when willow was used as fuel and around 38% with miscanthus. McIlveen-

Wright et al. [108] noticed in sensitivity analysis that in the best conditions around 84 $/MWh can be generated in the 250 kWe biomass SOFC system.

The most commonly used process simulators still remain HYSYS and Aspen Plus^{TM}, which predict the physical, thermodynamical and other properties of various materials used in chemical and mechanical engineering. Both software packages are quite similar and based on blocks corresponding to unit operations and chemical reactions, through which most of the operations can be simulated. A complete process flow sheet is reconstructed by interconnecting blocks using material, work and heat streams [115]. One of the benefits of using the flowsheeting software for SOFC combined system modelling is that analysis can be performed based on mass and energy balance equations in a scale process system. It is also possible to determine potential effects caused by changes of design parameters on the energetic performance of a SOFC combined system [110]. In addition, a parametric study usually is performed to specify SOFC operating temperature and fuel utilisation factors as well as to define the optimum operating conditions for individual components of a system, such as steam-to-carbon ratio, compressor pressure, specific work output, cycle with or without anode recirculation, etc. The software can also be used to perform economic analysis of various systems. Unfortunately, commercially available process simulators are not equipped with built-in units to simulate fuel cell operations. Zhang et al. [103] found that Aspen Plus^{TM} does not accommodate electrochemical reactions easily and also the stream mixing and transfer functions do not recognise ion transfer. Thus, two main system modelling approaches can be found in the literature.

In the first approach a SOFC stack model is usually defined in a programming language, such as C++, Fortran or Visual Basic and then it is linked to process simulation software packages [109-110, 116-118]. In this approach, the developed model of SOFC stack provides electrochemistry, momentum, heat and mass transfer inside the cell. Bo et al. [116] proposed a thermodynamic model for a tubular SOFC stack fed with natural gas with internal reforming, exhaust gas residual combustion and internal air preheating. The system consisted of an ejector, a heat exchanger, an anode and a cathode of the SOFC, afterburner and was modelled with Aspen Plus^{TM}. Similarly to the previous example, Suther et al. [118] developed a steady-state thermodynamic model for a hybrid solid oxide fuel cell – gas turbine cycle with a zero-dimensional macro-level SOFC model, which was integrated with the Aspen Plus model. Moreover, SOFC Fortran model by [118] was integrated into the hybrid SOFC- gas turbine cycle model in Aspen Plus™ process simulator by [110]. In this approach, when the SOFC model was executed within the hybrid cycle model, the flow rates and composition of the streams entering the SOFC were fed to the SOFC model based on the calculation of the hybrid model configuration and operating parameters in Aspen Plus™. Simulation results of a methane fuelled hybrid system showed that increasing the SOFC operating temperature increases the efficiency of the cycle, but decreases the system net specific work. The increase of fuel utilisation factor had a positive impact on the cycle net specific work [110]. Anderson at el. [111] included electrochemistry, diffusion phenomena and comprehensive reforming kinetics in a developed model for the methane fed SOFC using in-built HYSYS features. Three methods for SOFC modelling such as the equilibrium model, the two-stage PFTR model and the recycled reforming model were tested. It was found that only the last approach – the recycled reforming model provides reasonable results over a wide range of operating conditions and provides good consistency with experimental data.

In the second system modelling approach, the SOFC stack model is represented by Aspen Plus™ or HYSYS functions and unit operation modules [4, 8, 12, 103]. Zhang et al. [103] used a rather simple model without detailed consideration of cell voltage, diffusion phenomena and reforming kinetics. However, they applied the *Fsplitter* block available in the Aspen Plus™ to meet the desired S/C ratio value using *Design-spec* function [119]. Another *Design-spec* function was used to estimate the required inlet fresh pressure to drive recycling of the anode gas. Sensitivity analysis showed that the cell electrical efficiency reached a maximum value of 52% at the fuel utilisation equal to 0.85. The increase in fuel utilisation caused an increase in CO_2 concentration at the anode outlet stream [103]. In a study performed by [8] the cell voltage calculation was implemented in Aspen Plus™ by using a *Design-spec* Fortran block function. The model predicts cell voltage using semi-empirical correlations proposed by [6]. It also incorporates four additional semi-empirical equations to account for the effects of operating pressure and temperature, current density and fuel/air composition on the actual voltage [8]. In the study three reactions were considered in the anode block: electrochemical reaction (4.10), methane steam reforming reaction (4.11) and water-gas shift reaction (4.12), which were assumed using a simplified stoichiometric reactor block available in Aspen Plus™:

$$H_2 + \frac{1}{2}O_2 \rightarrow H_2O \tag{4.10}$$

$$CH_4 + H_2O \rightarrow CO + 3H_2 \qquad\qquad (4.11)$$

$$CO + H_2O \rightarrow CO_2 + H_2 \qquad\qquad (4.12)$$

Simulations results showed that if the fuel utilisation factor increased from 0.6 to 0.95, fuel cell voltage decreased due to the more fuel being depleted and polarisation loss at the anode. Voltage was reduced due to Ohmic and concentration losses if the utilisation factor increased to a magnitude larger than 0.85. It was concluded that the system should work with this value of the fuel utilization factor to obtain the maximum efficiency. Simulation results also indicated that the SOFC-GT cycle system was capable of achieving high electrical generation efficiency of 68.2%. A simplified SOFC stack model was used as well in the CPOx based micro-CHP system Aspen simulations operating with natural gas performed by [4] within the FP7 European Collaborative project FC-DISTRICT. The model utilised the Peng-Robinson property method. The air/fuel ratio in the CPOx reforming process was around 0.31, while the fuel utilisation in the SOFC stack was 75%. Heat and mass balance for the overall system was generated using unit operations available in the Aspen Plus^{TM} simulator. Simulation results enabled to evaluate system performance.

An excellent example of using unit operation blocks available in the commercial process simulator was presented by [105]. Indeed, Aspen Plus^{TM} is not capable to simulating cell half reactions (4.13) and (4.14), which take place at the cell level, therefore the overall reaction (4.15) has to be used to simulate the electrochemical reaction:

$$\frac{1}{2}O_2 + 2e^- \rightarrow O^{2-} \qquad\qquad \text{cathode half} - \text{cell reaction} \qquad (4.13)$$

$$H_2 + O^{2-} \rightarrow H_2O + 2e^- \qquad\qquad \text{anode half} - \text{cell reaction} \qquad (4.14)$$

$$H_2 + \frac{1}{2}O_2 \rightarrow H_2O \qquad\qquad \text{overall reaction} \qquad (4.15)$$

Thus, the anode is simulated using a Gibbs reactor module, while the cathode is represented by a separator module available in Aspen Plus™. The Nernst equation was used to calculate cell voltage. Ohmic, activation and concentration losses were subtracted from cell voltage using empirical and semi-empirical correlations. All these equations were specified in Aspen Plus™ using the *Design-spec* blocks. The input fuel flow rate was varied using the Aspen Plus™ design specification block to obtain the desired DC power [105].

Discussed numerical results demonstrated that current commercial process simulators provide a convenient way to perform detailed thermodynamic and parametric analysis of SOFC systems. They can also assess the effects of several operating condition variation on system performance. Therefore, it can be concluded, that if the goal is to improve the performance of a system by increasing its efficiency or output power, process simulators are appropriate tools to reach that goal.

The present challenge in multi-scale SOFC modelling is development of computationally efficient SOFC stack models for system operation, which will enable optimisation of system design. Their successful development and implementation would provide a significant boost for the commercialisation of CHP technologies based on SOFC.

5. Final remarks

Based on the literature review of SOFC systems it can be noticed that most numerical models were consisted of two modules, which can be used to simulate the behaviour of the SOFC stack and the remaining Balance of Plant (BoP) components. The SOFC stack model was usually based on the minimum Gibbs Free Energy approach, which is widely used in simplified models. In more advanced cases, the SOFC stack is represented by a 3D Computational Fluid Dynamics (CFD) model, which includes gas evolution into gas channels, porous electrodes and electrochemical fuel cell response. However, BoP components in the SOFC system model are represented by user defined units such as reformer, burner, heat exchanger, cooler, pump and compressor. The proper units can be defined using Fortran programming code or by those available in the commercial software packages such as ASPEN Plus or HYSYS and relations between them.

It seems that the major problem such as bridging the gap between models at different scales has already been solved. However, the problem of improving our understanding of SOFC electrochemical detailed kinetics has not, as yet, been successfully addressed.

References

[1] R. Bove, S. Ubertini S., *Modeling Solid Oxide Fuel Cells: Methods, Procedures and Techniques*, Springer, 2008, ISBN 978-1-4020-6995-6.

[2] J. Milewski, K. Świrski, M. Santarelli, P. Leone, *Advanced Methods of Solid Oxide Fuel Cell Modeling,* Springer, 2011, ISBN-10: 0857292617.

[3] B. Huang, Y. Qi, A. K. M. Murshed, *Dynamic Modelling and Predictive Control in Solid Oxide Fuel Cells: First Principle and Data-based Approaches*, Wiley, 2013, ISBN: 978-1-118-50103-0.

[4] D. Schimanke, O. Posdziech, B. E. Mai, S. Kluge, Th. Strohbach, Ch. Wunderlich, *ECS Trans*, **35**, 1 (2011) 231-242.

[5] A. D. Meadowcroft, S. Howroyd, K. Kendall, M. Kendall, *ECS Trans*, **57**, 1 (2013) 451-457.

[6] S. Campanari, *J. Power Sources*, **92** (2001) 26-34.

[7] J. Ugartemendia, J. X. Ostolaza, I. Zubia, *Energies*, **6** (2013) 5046-5068.

[8] M. Ameri, R. Mohammadi, *Intern. J. Energy Research*, **37** (2013) 412-425.

[9] W. G. Bessler, S. Gewies, C. Willich, G. Schiller, K. A. Friedrich, *Fuel Cells*, **10** (2010) 411-418.

[10] I. Frenzel, A. Loukou, D. Trimis, F. Schroeter, L. Mir, R. Marin, B. Egilegor, J. Manzanedo, G. Raju, M. de Bruijne, R. Wesseling, S. Fermandes, J. M. Ch. Pereira, G. Vourliotakis, M. Founti, O. Posdziech, *Energy Procedia*, **28** (2012) 170-181.

[11] I. Staffell, A review of small stationary fuel cell performance, 2009. http://www.academia.edu/1073987/A_review_of_small_stationary_fuel_cell_perfor mance.

[12] B. Zakrzewska, Z. Jaworski, *Computers & Chemical Engineering*, **35** (2011) 434-445.

[13] W. G. Bessler, *ECS Transactions*, **35** (2011) 859-869.

[14] M. Andersson, J. Yuan, B. Sunden, *Applied Energy*, **87** (2010) 1461-1476.

[15] K. N. Grew, W. K. S. Chiu, *J. Power Sources*, **199** (2012) 1-13.

[16] S. Cordiner, M. Feola, V. Mulone, F. Romanelli, Analysis of a SOFC energy generation system fuelled with biomass reformate, *Energy: production, distribution and conservation*, Milan 2006, pp. 103-119.

[17] W. G. Bessler, J. Warnatz, D. G. Goodwin, *Solid State Ionics*, **177** (2007) 3371-3383.

[18] D. Cui, Q. Liu, F. Chen, *J. Power Sources*, **195** (2010) 4160-4167.

[19] F. Bidrawn, R. Kungas, J. M. Vohs, R. J. Gorte, *J. Electrochemical Society*, **158**, 5 (2011) B514-B525.

[20] C. R. Kreller, M. E. Drake, S. B. Adler, H. Y. Chen, H. C. Yu, K. Thorton, J. R. Wilson, S. A. Barnett, *ECS Trans*, **35**, 1 (2011) 815-822.

[21] V. Yurkiv, A. Latz, W. G. Bessler, *ECS Transactions*, **57**, 1 (2013) 2637-2647.

[22] J. R. Wilson, W. Kobsiriphat, R. Mendoza, H. Y. Chen, T. Hines, J. M. Hiller, D. J. Miller, K. Thornton, P. W. Voorhees, H. Mizusaki, *Solid Oxide Fuel Cells 10 (SOFC-X)*, The Electrochemical Society, Nara, Japan, 2007, pp.1879-1887.

[23] J. J. R. Izzo, A. S. Joshi, K. N. Grew, W. K. S. Chiu, A. Tkachuk, S. H. Wang, W. Yun, *J. Electrochemical Society,* **155**, 5 (2008) B504-B508.

[24] P. R. Shearing, J. Golbert, R. J. Chater, N. P. Brandon, *Chem. Eng. Sci.*, **64** (2009) 3928-3933.

[25] H. Iwai, N. Shikazono, T. Matsui, H. Teshima, M. Kishimoto, R. Kishida, D. Hayashi, K. Matsuzaki, D. Kanno, M. Saito, H. Muroyama, K. Eguchi, N. Kasagi, H. Yoshida, *J. Power Sources*, **195** (2010) 955-961.

[26] S. B. Adler, *Chemical Reviews*, **104** (2004) 4791-4843.

[27] J. Fleig, *J. Power Sources*, **105** (2002) 228-238.

[28] Y. X. Lu, C. Kreller, S. B. Adler, *J. Electrochemical Society*, **156** (2009) B513-B252.

[29] H. Zhu, R. J. Kee, V. M. Janardhanan, O. Deutschmann, D. G. Goodwin, *J. Electrochemical Society*, **152**, 12 (2005) A2427-A24440.

[30] V. M. Janardhanan, O. Deutschmann, *Chem. Eng. Sci.*, **62** (2007) 5473-5486.

[31] V. M. Janardhanan, V. Heuveline, O. Deutschmann, *J. Power Sources*, **178** (2008) 368-372.

[32] M. Vogler, A. Bieberle-Hutter, L. Gauckler, J. Warnatz, W. G. Bessler, *J. Electrochemical Society*, **156**, 5 (2009) B663-B6727.

[33] M. Ni, M. K. H. Leung, D. Y. C. Leung, *J. Power Sources*, **168**, 2 (2007) 369-378.

[34] J. Shi, X. Xue, *Electrochimica Acta*, **55**, 18 (2010) 5263-5273.

[35] M. Andersson, J. Yuan, B. Sunden, *Intern. J. Heat and Mass Transfer*, **55** (2012) 773-788.

[36] M. M. Hussain, X. Li, I. Dincer, *Intern. J. Hydrogen Energy*, **34** (2009) 3134-3144.

[37] S. Hosseini, K. Ahmed, M. O. Tade, *J. Power Sources*, **234** (2013) 180-196.

[38] D. Goodwin, Cantera: an object-oriented software toolkit for chemical kinetics, thermodynamics and transport processes, Caltech, Pasadena, http://code.google.com/p/cantera.

[39] W. G. Bessler, *Solid State Ionics*, **176** (2005) 997-1011.

[40] L. C. R. Schneider, C. L. Martin, Y. Bultel, L. Dessemond, D. Bouvard, *Electrochimi. Acta*, **52** (2007) 3190-3198.

[41] B. Kenney, M. Valdmanis, C. Baker, J. G. Pharoah, K. Karan, *J. Power Sources*, **189**, 2 (2009) 1051-1059.

[42] J. Sanyal, G. H. Goldin, H. Zhu, R. J. Kee, *J. Power Sources*, **195** (2010) 6671-6679.

[43] A. Abbaspour, J. L. Luo, K. Nandakumar, *Electrochimica Acta*, **55** (2010) 3944-3950.

[44] A. Bertei, B. Nucci, C. Nicolella, *Chem. Eng. Sci.*, **101** (2013) 175-190.

[45] A. Bertei, H. W. Choi, J. G. Pharoah, C. Nicolella, *Powder Technology*, **231** (2012) 44-53. [46] G. M. Goldin, H. Zhu, R. J. Kee, D. Bierschenk, S.A. Barnett, *J. Power Sources*, **187** (2009) 123-135.

[47] Z. Qu, P. V. Aravin, S. Z. Boksteen, N. J. J. Dekker, A. H. H. Janssen, N. Woudstra, A. H. M. Verkooijen, *Intern. J. Hydrogen Energy*, **30** (2011) 10209-10220.

[48] P. Fan, G. Li, Y. Zeng, X. Zhang, *Intern. J. Thermal Sciences*, **77** (2014) 1-10.

[49] T. X. Ho, P. Kosinski, A. C. Hoffmann, A. Vik, *Chem. Eng. Sci.*, **64** (2009) 3000-3009.

[50] M. Peksen, *Intern. J. Hydrogen Energy*, **39** (2014) 5137-5147.

[51] V. Novaresio, M. Garcia-Camprubi, S. Izquierdo, P. Asinari, N. Fueyo, *Computer Physics Communications*, **183** (2012) 125-146.

[52] M. Garcia-Camprubi, S. Izquierdo, N. Fueyo, *Renewable and Sustainable Energy Review*, **33** (2014) 701-718.

[53] S. A. Hajimolana, M. A. Hussain, W. M. Ashri Wan Daud, M. Soroush, A. Shamiri, *Renewable and Sustainable Energy Reviews*, **15** (2011) 1893-1917.

[54] A. Pramuanjaroenkij, S. Kakac, X. Y. Zhou, *Intern. J. Hydrogen Energy*, **33** (2008) 2547-2565.

[55] T. X. Ho, *Intern. J. Hydrogen Energy*, **39** (2014) 6680-6688.

[56] W. Li, Y. Shi, Y. Luo, N. Cai, *Intern. J Hydrogen Energy*, **XXX** (2014) 1-13.

[57] R. B. Bird, W. E. Stewart, E. N. Lighfoot, Transport Phenomena, Wiley, New York, 2002.

[58] D. L. Damm, A. G. Fedorov, *J. Power Sources*, **159** (2006) 1153-1157.

[59] D. L. Damm, A. G. Federov, *J. Fuel Cell Science and Technology*, **2** (2005) 258-262.

[60] D. L. Damm, A. G. Fedorov, *J. Power Sources*, **143** (2005) 158-165.

[61] K. J. Daun, S. B. Beale, F. Liu, G. J. Smallwood, *J. Power Sources*, **157** (2006) 302-310.

[62] X. Zhang, G. Li, J. Li, Z. Feng, *Energy Conversion and Management*, **48**, 3 (2007) 977-989.

[63] S. Grosso, L. Repetto, P. Costamagna, IP-SOFC model, Chapter 6: *Modeling Solid Oxide Fuel Cells, Methods, Procedures and Techniques, Fuel Cells and Hydrogen Energy,* Editors: R. Bove, S. Ubertini, Springer, 2008. E-ISBN-13:978-1-4020-6995-6.

[64] P. Costamagna, K. Honegger, *J. Electrochemical Society*, **145**, 11 (1998) 3995-4007.

[65] W. G. Bessler, S. Gewies, M. Vogler, *Electrochimica Acta*, **53** (2007) 1782-1800.

[66] K. Wang, D. Hissel, M. C. Pera, N. Steiner, D. Marra, M. Sorrentino, C. Pianese, M. Monteverde, P. Cardone, J. Saarinen, *Intern. J. Hydrogen Energy*, **36** (2011) 7212-7228.

[67] J. Arriagada, P. Olausson, A. Selimovic, *J. Power Sources*, **112** (2002) 54-60.

[68] J. Milewski, K. Swirski, *J. Hydrogen Energy,* **34**, 13 (2009) 5546-5553.

[69] J. Nielsen, J. Hjelm, *Electrochimica Acta*, **115** (2014) 31-45.

[70] H. Mounir, M. Belaiche, A. El Marjani, A. El Gharad, *Energy*, **66** (2014) 378-386.

[71] S. S. Wei, T. H. Wang, J. D. Wu, *Energy*, **XXX** (2014) 1-9.

[72] A. Nakajo, Ch. Stiller, G. Harkegard, O. Bolland, *J. Power Sources,* **158** (2006) 287-294.

[73] L. K. Chiang, H. Ch. Liu, Y. H. Shiu, Ch. H. Lee, R. Y. Lee, *Renewable Energy,* **33** (2008) 2580-2588.

[74] A. Selimovic, M. Kemm, T. Torisson, M. Assadi, *J. Power Sources,* **145** (2005) 463-469. [75] M. F. Serincan, U. Pasaogullari, N. M. Sammes, *J. Power Sources*, **195** (2010) 4905-4914.

[76] V. Lawlor, S. Griesser, G. Buchinger, A. G. Olabi, S. Cordiner, D. Meissner, *J. Power Sources,* **193** (2009) 387-399.

[77] A. F. Mueller, J. Brouwer, J. Van herle, D. Favrat, *Intern. J. Hydrogen Energy*, **37** (2012) 9269-9286.

[78] M. Peksen, A. Al-Masri, L. Blum, D. Stolten, *Intern. J. Hydrogen Energy*, **38** (2013) 4099-4107.

[79] J. Wei., Models Development and System Simulation of Solid Oxide Fuel Cell/gas, UMI Microform 3232512, chapter 4 (2006) 4-28.

[80] T. Ota, M. Koyama, Ch. J. Wen, K. Yamada, H. Takahashi, *J. Power Sources*, **118** (2003) 430-439.

[81] J. Qu, A. Fedorov, C. Haynes, *An integrated approach to modelling and mitigating SOFC failure*, Georgia Tech, Atlanta, GA 30332-0405, Final Technical Report, 1 July 2006. DOE: DE-FC26-02NT41571.

[82] T. Tanaka, Y. Inui, A. Urata, T. Kanno, *Energy Conversion and Management*, **48** (2007) 1491-1498.

[83] J. D. J. VanderSteen, J. G. Pharoah. *The effect of radiation heat transfer in Solid Oxide Fuel Cell modelling*, Combustion Institute, Canadian Section, Spring Technical Meeting, Queen's University, 9-12 May 2004.

[84] J. Hartvigsen, S. Elangovan, A. Khandkar, *Modeling, design and performance of Solid Oxide Fuel Cells*, 105, Prof. of Zirconia V., 2002.

[85] H. Yakabe, T. Ogiwara, I. Yasuda, M. Hishunuma, *J. Power Sources*, **102** (2001) 144-154. [86] M. Lockett, M. J. H. Simmons, K. Kendall, *J. Power Sources*, **131** (2004) 243-246.

[87] S. Murthy, A. G. Fedorov, *J. Power Sources*, **124** (2003) 453-458.

[88] K. J. Kattke, R. J. Braun, A. M. Colclasure, G. Goldin, *J. Power Sources*, **196** (2011) 3790-3802.

[89] S. K. Mazumder, K. Acharya, C. L. Haynes, R. Williams, M. R. von Spakovsky, D. J. Nelson, D. F. Rancruel, J. Hartvigsen, R. S. Gemmen, *IEEE Transactions on Power Electronics*, **19**, 5 (2004) 1263-1278.

[90] A. Nakajo, P. Tanasini, S. Diethelm, J. Van herle, D. Favrat, *J. Electrochemical Society,* **158**, 9 (2011) B1102-B1118.

[91] O. Posdziech, System concepts & BoP components, Joint European Summer School for Fuel Cell and Hydrogen Technology, Viterbo, 29th August – 2nd September 2011.

[92] Z. Al-Hamamre, S. Vob, D. Trimis, *Intern. J. Hydrogen Energy,* **34** (2009) 827-832.

[93] A. Al-Musa, S. Shabunya, V. Martynenko, S. Al-Mayman, V. Kalinin, M. Al-Juhani, K. Al-Enazy, *J. Power Sources*, **246** (2014) 473-481.

[94] W. Wang, Y. Cao., *Intern. J. Hydrogen Energy,* **35**, 24 (2010) 13280-13289.

[95] O. Posdziech, B. E. Mai, C. Wunderlich C., Voss S., Status and market opportunities of solid oxide fuel cell based cogeneration systems, International Gas Union Research Conference, (2011) 1-17.

[96] V. Liso, A. Ch. Olesen, M. P. Nielsen, S. K. Kaer, *Energy,* **36** (2011) 4216-4226.

[97] N. Laosiripojana, W. Wiyaratn, W. Kiatkittipong, A. Arpornwichanop, A. Soottitantawat, S. Assabumrungrat, *Engineering Journal,* **13**, 31 (2009) 65-83.

[98] C. M. Finnerty, N. J. Coe, R. H. Cunningham, R. M. Ormerod, *Catalysis Today,* **46** (1998) 137-145.

[99] C. M. Finnerty, R. M. Ormerod, *J. Power Sources,* **86** (2000) 390-394.

[100] J. Kupecki, J. Jewulski, K. Badyga, *Rynek Energii,* **6**, 97 (2011) 157-162.

[101] T. S. Lee, J. N. Chung, Y. Ch. Chen, *Energy Conversion and Management,* **52** (2011) 3214-3226.

[102] J. Kupecki, J. Milewski, J. Jewulski, *Cent. Eur. J. Chem.* **11**, 5 (2013) 664-671.

[103] W. Zhang, E. Croiset, P. L. Douglas, M. W. Fowler, E. Entchev, *Energy Conversion and Management,* **46** (2005) 181-196.

[104] G. Bae, J. Bae, P. Kim-Lohsoontorn, J. Jeong, *Intern. J. Hydrogen Energy,* **35** (2010) 12346-12358.

[105] T. Tanim, D. J. Bayless, J. P. Trembly, *J. Power Sources,* **245** (2014) 986-997.

[106] A. Ersoz, A. Sarioglan, S. Ozdogan, Investigation of an integrated autothermal reforming and SOFC micro cogeneration system for power generation, WHEC, 16, 13-16 June 2006, Lyon, France.

[107] M. Rokni, *Energy,* **XXX** (2014) 1-13.

[108] D. R. McIlveen-Wright, M. Moglie, S. Rezvani, Y. Huang, *Intern. J. Energy Research,* **35** (2011) 1037-1047.

[109] F. Zabihian, A. S. Fung, *Energy Conversion and Management,* **76** (2013) 571-580.

[110] F. Zabihian, A. S. Fung, *Sustainable Energy Technologies and Assessments,* **6** (2014) 51-59.

[111] T. Anderson, P. Vijay, M. O. Tade, *Chem. Eng. Research and Design,* **92** (2014) 295-307.

[112] J. Kupecki, J. Jewulski, *Multi-level mathematical modelling of Solid Oxide Fuel Cells,* Clean Energy for better environment, chapter **4** (2012) 53.

[113] J. Milewski, A. Miller, *Mathematical model of SOFC for power plant simulations*, Proceedings of ASME Turbo EXPO 2004, 14-17 June, Vienna, Austria, GT2004-53787.

[114] B. Elmegaard, N. Houbak, *DNA – A general energy system simulation tool*, In: Proceeding of SIMS, 2005, Trondheim, Norway.

[115] R. Schefflan, Teach yourself the basics of Aspen Plus™, AIChE, Wiley, A John Wiley & Sons, Inc., 2011, New Jersey. ISBN 978-0-470-56795-1.

[116] Ch. Bo, Ch. Yuan, X. Zhao, C. B. Wu, M. Q. Li, *Clean Techn. Environ. Policy*, **11** (2009) 391-399.

[117] E. Riensche, E. Achenbach, D. Froning, M. R. Haines, W. K. Heidug, A. Lokurlu, S. von Andrian, *J. Power Sources*, **86** (2000) 404-410.

[118] T. Suther, A. S. Fung, M. Koksal, F. Zabihian, *Intern. J. Energy Research,* **35,** 7 (2011) 616-632.

[119] Aspen Plus™ Users Guide, 10.2, 2000. Aspen Tech Ltd., Cambridge MA, USA.

Acknowledgments

The research program leading to these results received funding from the European Union's Seventh Framework Programme (FP7/2007-2013) for the Fuel Cells and Hydrogen Joint Technology Initiative under grant agreement no [303415]. Information contained in the paper reflects only view of the authors. The FCH JU and the Union are not liable for any use that may be made of the information contained therein. Acknowledgments are due to the partners of the SAPIENS project.

The work was also financed from the Polish research funds awarded for the project No. 2750/7.PR/2013/2 of international cooperation within SAPIENS in years 2012-2015.